Virus
and the
Whale

Exploring Evolution
in Creatures
Small and Large

Virus
and the
Whale

Exploring Evolution in Creatures Small and Large

Judy Diamond, *Editor*

With Carl Zimmer, E. Margaret Evans,
Linda Allison, and Sarah Disbrow

NATIONAL SCIENCE TEACHERS ASSOCIATION
Arlington, Virginia

Claire Reinburg, Director
Judy Cusick, Senior Editor
J. Andrew Cocke, Associate Editor
Betty Smith, Associate Editor
Robin Allan, Book Acquisitions Coordinator

PRINTING AND PRODUCTION, Catherine Lorrain-Hale, Director
Nguyet Tran, Assistant Production Manager
Jack Parker, Desktop Publishing Specialist

Linda Olliver, Cover and Interior Design
Angie Fox, Cover Illustration

NATIONAL SCIENCE TEACHERS ASSOCIATION
Gerald F. Wheeler, Executive Director
David Beacom, Publisher

08   07   06   4   3   2   1

Library of Congress Cataloging-in-Publication Data

Virus and the whale : exploring evolution in creatures small and large / Judy Diamond, editor ; with Carl Zimmer ... [et al.].
     p. cm.
 Includes bibliographical references.
 ISBN 0-87355-263-6
 1. Evolution (Biology)--Study and teaching (Middle school) 2. Evolution (Biology)--Study and teaching (Middle school)--Activity programs. I. Diamond, Judy. II. Zimmer, Carl, 1966-

 QH362.V57 2006
 576.8071'2--dc22
                                                 2005024464

Contents

Preface

Judy Diamond
Director
Explore Evolution project

For the past two decades, my family and I have conducted field research on the behavior and evolution of two species of parrots from New Zealand (Diamond and Bond 1999). Over the years, we compared and contrasted the behavior of these closely related species to see where common patterns reinforced common ancestry, or where different patterns suggested adaptations to diverse environments.

As young field researchers, my children understood that evolution was part of how our family viewed the natural world. However, each time they returned from the field to their classrooms in Lincoln, Nebraska, they encountered strong denials of our ideas. Many of their classmates dismissed any possibility that evolutionary ideas were correct or relevant. Our children often asked us how we, as parents, could be so sure that our ideas were right, and how so many other children and their parents could be so wrong.

The Explore Evolution project started as a way to help young people and adults gain some of the experience of what it is like to study evolution as a scientist. Most biological scientists view the world from an evolutionary perspective. The geneticist Theodosius Dobzhansky said in 1964, "Nothing in biology makes sense except in the light of evolution." This is as true today as it was 40 years ago (Dobzhansky 1964).

In biology, thinking like a scientist is, in large part, thinking about evolution. As educators, our task is to help young people feel what it means to view the world from a scientific and evolutionary perspective. This book invites you to engage youth in activities that are based on current research projects that have had a major influence on how scientists today think about evolution.

Designed to be used with middle-school-age youth, the activities in this book work in almost any setting, they are youth-centered, and they are designed to encourage social interaction. Each of the seven activities incorporates concepts of inquiry-based learning and the 4-H Youth Development experiential learning model. Along with the activities, the book provides many resources for teachers and youth leaders. The first section presents an overview of what you need to know about how to use this book with kids. It includes learning outcomes for each activity tied to the National Science Education Standards, assessment questions, and materials needed. The first chapter, by Carl Zimmer, gives an introduction to evolution for those who want for more background on the topic. The second chapter, also by Carl

Zimmer, gives background information on the evolution of each of the seven organisms that are investigated in this book. The third chapter, by Margaret Evans, helps introduce teachers and youth leaders to common ways that children and adults think and learn about evolution.

This book is a product of the Explore Evolution project, a partnership forged between science museums and 4-H organizations in the Midwest to bring current research on evolution to middle school youth, educators, and the general public. Our museum partners include the Exhibit Museum of Natural History of the University of Michigan, the Kansas Museum and Biodiversity Center at the University of Kansas, the Sam Noble Oklahoma Museum of Natural History at the University of Oklahoma, the Texas Memorial Museum at the University of Texas at Austin, the University of Nebraska State Museum and the Science Museum of Minnesota. Our 4-H partners are the statewide 4-H Youth Development programs in Iowa, Minnesota, Nebraska, Texas, and Wyoming. The Explore Evolution project has produced new permanent exhibit galleries for the partner museums. This book is designed in parallel with the galleries—to be used independently or in conjunction with the exhibits. The exhibits and the book focus on the same seven current and influential evolutionary research projects.

The central theme of Explore Evolution is to show common patterns and principles in the evolution of organisms ranging in size from viruses to whales. The secondary theme is to show how scientific research uncovers new ideas about how the natural world operates, and how these ideas are continuously modified by new research. Overall, we emphasize that evolution is the central unifying principle for a scientific understanding of the natural world. The book provides durable, hands-on and easy-to-use activities to bring investigations about evolution to youth in any of the settings where they can do science with their peers.

Acknowledgments

This book represents a huge effort on the part of many people. Foremost, we are indebted to the scientists who provided us with the fascinating research projects that are the subject of this book, and who also worked with us wholeheartedly throughout this enormous undertaking. For this, we thank Cameron R. Currie, University of Wisconsin; Sherilyn C. Fritz, University of Nebraska-Lincoln; Philip D. Gingerich, University of Michigan; B. Rosemary and Peter R. Grant, Princeton University; Henrik Kaessmann, University of Lausanne, Switzerland; Kenneth Y. Kaneshiro, University of Hawaii at Manoa; Svante Pääbo, Max Planck Institute for Evolutionary Anthropology; Edward C. Theriot, University of Texas at Austin; and Charles Wood, University of Nebraska-Lincoln.

We also recognize the support given to us by the museums that participated as partners in the Explore Evolution project. For their commitment to the project, we acknowledge and thank museum directors Ellen J. Censky of the Sam Noble Oklahoma Museum of Natural History at the University of Oklahoma, Priscilla Grew of the University of Nebraska State Museum, Amy Harris of the Exhibit Museum of Natural History at the University of Michigan, Eric Jolly of the Science Museum of Minnesota, Leonard Krishtalka of the Natural History Museum and Biodiversity Research Center at the University of Kansas, and Edward Theriot of the Texas Memorial Museum at the University of Texas at Austin.

Many museum staff members played significant roles in making this book possible. Foremost, we thank Angie Fox for her outstanding illustrations, which appear throughout in the book and the exhibit, and for her help in producing the book graphics. We also thank Katrina Hase, Rob Sharot, Roger Barrett, Paul Martin, David Chittenden, and many others from the staff of the Science Museum of Minnesota who played a central role in developing the exhibit galleries. Debra Meier, Joel Nielson, and Ron Pike (Nebraska), John Klausmeyer (Michigan), Brad Kemp and Bruce Sherling (Kansas), and John Maisano (Texas) contributed valuable expertise to the project and assisted in the development of exhibits and in the overall organization of the Explore Evolution project.

Several individuals deserve special mention for the assistance they provided to the project. We greatly appreciate the advice of John West for his critical help in the virology activity and exhibits. We also thank Lisa Diamond of the Stanford Genome Technology Center for her valuable help in developing the human-chimp gene comparison graphic. Finally, we thank Judy Scotchmoor of the University of California Museum of Paleontology for her unfailing support and many useful suggestions.

We are also indebted to the efforts of many individuals whose work greatly enhanced the quality and usefulness of this book. The Explore Evolution Nebraska outreach team, coordinated by Kathy French, and assisted by Cindy Loope, Eileen Cunningham, Estella Wolf, and Ann Cusick, trial tested the book with middle school children at several stages of its development and gave us much valuable advice. We also thank the 38 middle school kids who participated in the trial testing and gave us such valuable feedback.

The Explore Evolution evaluation team, coordinated by Amy Spiegel, Wendy Gram, and Margaret Evans, and assisted by Deborah Kay, Cindy Loope, Brandy Frazier, Sarah Cover, and Medha Tare, provided valuable feedback and guidance throughout the development of the book and exhibit. Many 4-H leaders and museum educators have advised us throughout the course of this project and have played an important role in disseminating this project to youth leaders, teachers, and children. We thank all of them, and, in particular, Bradley Barker, Elizabeth Birnstihl, Stephen Carlson, Catherine Denison, Dianne Folkerth, Gene Gade, Lilliane Goeders, Teresa MacDonald, Robert Meduna, and Jay Staker.

We owe a special thanks to our partners and children, Alan, Benjamin, and Rachel; David, Anna, and Vanessa; Gus Buchtel; and William Wells, for supporting us during this project.

For guiding the project through the many stages from manuscript to publication, our thanks go to our editor Claire Reinburg, project editor Andrew Cocke, graphic designer Linda Olliver, and production director Catherine Lorrain-Hale at the NSTA Press.

Finally, we would like to thank the Informal Science Education program of the National Science Foundation (NSF). This material is based upon work supported by the NSF under Grant No. 0229294. Any opinions, findings, and conclusions or recommendations expressed in this material are those of the authors and do not necessarily reflect the views of NSF.

References

Diamond, J., and A. Bond. 1999. *Kea, bird of paradox: The evolution and behavior of a New Zealand parrot.* Berkeley: University of California Press.

Dobzhansky, T. 1964. Biology, molecular and organismic. *American Zoologist* 4: 449.

Introduction

Eugenie C. Scott
Executive Director
National Center for Science Education, Inc.

few years ago a publishing house editor told me I should write a book about my one-of-a-kind job as executive director of the National Center for Science Education (NCSE). NCSE defends the teaching of evolution in the public schools and opposes the teaching of religious views as science. I get to do a lot of public speaking and writing about the creationism/evolution controversy, and some radio and television work, which is fun. Mostly I work with my colleagues to advise people around the country on what to do when the issue of teaching evolution becomes controversial in a community or state. She said I should title the book, "Evangelist for Evolution."

I blanched.

What came to mind was an exchange with Duane Gish, debater extraordinaire of the Institute for Creation Research, on a radio call-in show. Duane said (as he had said many times before) "evolution is just as religious as creationism." I was sufficiently exasperated at yet again hearing this old saw to snap, "If evolution is a religion, it's a damned uninspiring one." And indeed, even though the scientific establishment concurs that living things are related through common ancestry, and even though university science libraries are laden with books and journals matter-of-factly discussing evolution, this central scientific idea does not inspire people to defend it with any where near the zeal of people attacking it. (We are especially aware of this when we compare the, ahem, finances of NCSE to that of the several major creationist organizations!)

Well, I didn't write Evangelist for Evolution—I wrote a different book (Scott 2004). Yet I think maybe someday I will, though not because evolution is a religion. Evolution is not a religion. I repeat. Evolution is not a religion. There is no "church of Darwin," evolutionists do not have rituals (not even a secret handshake), and no one is going to get salvation from "believing" in natural selection (or rejecting it). (If you don't know what natural selection is, hang in there; you'll read about it later in this book.) But religion is why evolution is viewed as controversial by so many Americans. A Gallup poll found that even though only 5% of scientists agreed with the statement "Humans were created by God in their present form about 10,000 years ago," 47% of the general public agreed (Witham 1997). I believe that the reason for the large percentage of rejection of evolution in the United States (though interestingly, not in other developed countries) is that people confuse evolution with atheism: they believe that they have to choose between evolution and God, between science and religion.

Unfortunately, Americans are not only illiterate in science, they are theological illiterates as well. Most are unaware that although some religious traditions (such as biblical literalism) are incompatible with evolution, literalism is not the dominant tradition in either Christianity or Judaism, the two largest faiths in the United States. Catholic parochial schools routinely teach evolution; Catholic and mainstream Protestant theologies reflect varieties of "theistic evolution:" that God created through the process of evolution. They do not believe in Genesis literalist special creationism, which is that God created everything all at one time, in its present form (and usually only about 10,000 years ago or less). Alas, there are many Catholics and mainstream Protestants who don't know their denomination's theology! But as an educator, you will need to deal with these misunderstandings. If a visitor says to you, "I'm a creationist" you might want to ask, "What kind?" because to many people the word *creationist* means that God created, on which all Christians, Jews and Muslims agree. Identifying as a creationist doesn't necessarily identify one as a special creationist: Someone describing him or herself as a creationist might be a theistic evolutionist. Pope John Paul, for example, was one of these creationists (he believed God created) who also accepted evolution. The confusion of terminology about who is a creationist only adds to other confusions about evolution and religion, but at least it is fairly easy to clear up.

The tough issues, though, go beyond definitions and concern God's involvement in creation. Is God guiding evolution? Is there a goal to evolution? Is there an ultimate purpose to the universe? If there are answers to these questions, they won't come from science, but from theology. Because science explains using only natural cause, it can't be used to test supernatural explanations. Obviously, an omnipotent power can create any way it wants, including all at once, or gradually through evolution—or create all at once and make it look like evolution happened. Any explanation is compatible with an omnipotent Creator, so therefore creationist claims cannot be tested. If a statement can't be tested against the natural world, it isn't a scientific statement. Using science, one cannot say whether God did or did not create, nor can the methods of science offer guidance about whether God is the ultimate cause of everything. Science is a limited way of knowing, limited to explaining only the natural world, and restricted to natural causes.

But as the philosopher once said, "To say nothing of God is not to say that God is nothing." If science explains things without mentioning God, that doesn't mean that science is saying, "God had nothing to do with it." It's funny: No one accuses science of ignoring God when we try to explain cell division. If a scientist says, "Enzyme X causes the cell to begin to divide, and enzyme Y causes spindle fibers to form, and enzyme Z causes the chromosomes to line up in the middle," no one castigates him/her for not mentioning "because God wanted the chromosomes to line up in the middle." We don't expect scientists to bring up God while explaining cell division—so why do some people get upset when we don't bring up God when talking about evolution? But people do get upset, and you will need to help them understand why such explanations are outside of science, and that indeed—unless they are special creationists—science is not challenging their religious views. Indeed, to say nothing of God is not to say that God is nothing.

What science *can* tell us, however, is that the evidence is overwhelming that species have a genealogical relationship, and that the pattern of evolution seen through the fossil record does not appear to be heading toward any particular goal. In the Miocene, we see a variety of horse adaptations: some browsers, some grazers, some with three toes, and some with one toe. Viewing those different adaptations used by horses 20 million years ago doesn't allow us to predict that one million years ago in the Pleistocene, only one-toed grazers would be left. Did God intend that *Equus* (the genus that includes the modern horse) would be the "winner" in the evolutionary game? Science can't say. Out of all the ape species of the Pliocene, did God intend a bipedal one with an opposable thumb to eventually give rise to *Homo sapiens*? Science can't say. These are theological questions, not scientific ones. Science describes what happened, when, and where; only philosophy and religion can speculate as to the ultimate meaning or ultimate cause of what happened.

This means religious individuals are free to hold most beliefs about God's involvement with evolution if they choose; there is nothing in the scientific understanding of evolution that prohibits someone from believing that God is the ultimate cause of everything in the universe, or even that God could be guiding evolutionary events. But it is not possible to verify such statements through science.

As you read the following book, I think you will agree that evolution is an exciting scientific idea. Students should know about it, for the sheer intellectual joy of the experience. I remember when I was in high school, my biology teacher, a kindly old gray-haired man named Mr. Rasmussen, did not teach us evolution, and I don't recall it being in the textbook. One day after school a friend and I were talking to him and one of us asked him about why there were so many different kinds of animals. He said, "Well, some people believe that some animals are better able to live in an environment than others, and they have more offspring and that kind comes to dominate in that environment, and the population gradually becomes different through time."

My head reeled. This was such a wonderful, simple explanation, and it made so much sense (it is the essence of Darwinian natural selection, a prime mechanism of evolution). I remember becoming terribly excited, and brimming over with questions. I wanted to know everything—did this explain why we no longer had dinosaurs around? Did people change like this? Before I could say anything, though, Mr. Rasmussen quickly added, "Of course, some people think that God created all the animals like they look today."

I hesitated. It sounded like I was being given the choice between science and religion, and I didn't want to sound like I was being antireligious. But so many questions wanted to tumble out! But Mr. Rasmussen held a finger to his lips, conspiratorially, and said, "Shhhh-hhh," as he eased us out of the classroom and closed the door.

That was the end of my evolution instruction until college, and even today, more than 40 years later, many students are similarly denied the excitement of learning this simple but elegant theory. I hope you, as teachers, youth leaders, or museum curators, can help people begin to appreciate evolution, and how it helps us make sense of so much of the living world. It is an important job and a very worthwhile one—and it will be made enjoyable by the activities you will be leading the children through.

Because even though evolution is not a religion, it is sorely in need of someone to tell people about it. At NCSE, we try to help people understand evolution—why they need not be afraid of it, and why their kids should be taught it in school. You also have the opportunity to help students and others to understand this quite astonishing idea that explains so much about nature. You don't have to be an evangelist for evolution, but you can certainly open up what for many people will be a new area of knowledge which is likely to lead to some profound understandings of living things.

I wish you success—and the enjoyment of sharing exciting scientific ideas!

References

Scott, E. 2004. *Creationism vs. evolution: An introduction.* Westport, CT: Greenwood.

Witham, Larry. 1997. Many scientists see God's hand in evolution. *Washington Times*, April 11, 1997, A8.

How to Use This Book

Judy Diamond

This book focuses on how scientists conduct research on the evolution of organisms ranging from a tiny virus to a huge whale. It features activities and background resources that explore seven current evolution research projects. The activities are designed to be used with middle-school-age youth in almost any setting ranging from classrooms to clubs and after-school programs. The activities can be used individually, in any order, and they can supplement text-based curricula. The activities are youth-centered and encourage social interaction: Youth proceed through the activity sections in small groups that work at their own pace. Background resources supplement the activities by providing teachers and youth leaders with an overview of evolutionary principles, along with a resource list of books and websites where more information is available.

The activities in this book require minimal preparation. This section provides three overviews to help you prepare to use the activities: 1) learning outcomes for each chapter tied to the relevant National Science Education Standards identified by the National Research Council of the National Academy of Sciences, 2) a summary of how the learning outcomes relate to the assessment questions at the end of each activity, and 3) the materials required for each activity.

Three chapters in the front of the book provide teachers and youth leaders with background on the topics being covered. In Chapter one, science writer Carl Zimmer gives a general overview about evolution and its major themes: variation, inheritance, selection, and time. Chapter two, also by Carl Zimmer, gives specific background on the evolution of the seven organisms that are investigated in the activities. Chapter three, by cognitive psychologist Margaret Evans, describes how the seven activity chapters incorporate concepts of inquiry-based learning and the 4-H Youth Development experiential learning model. This chapter also shows how research on learning can help teachers and youth leaders recognize the ways that children and adults think about evolution.

The second half of the book comprises the seven activities. Each activity begins with a short essay written for youth that gives an overview of a scientist's research project on a particular organism. This introductory essay is followed by a four-part hands-on activity. The first part is a brief exercise that introduces users to the organism being studied. Two sections of investigations follow, each based on an aspect of the scientist's research. The fourth and final section is an assessment tool in which youth are asked to think like a science reporter and summarize their understanding of the activities in that chapter.

National Science Education Standards

Each of the activities included in this book is based on the National Science Education Content Standards C and D for grades 5–8 in Life Science and Earth and Space Science (NRC 1996). See Table 1, on the next page, for a complete listing of the learning outcomes for each activity tied to the relevant content standard.

Reference

National Research Council (NRC). 1996. *National science education standards.* Washington, DC: National Academy Press.

Table 1: Key to the National Science Education Standards Relevant to Each Explore Evolution Activity.

ACTIVITY NUMBER	ACTIVITY TITLE	LEARNING OUTCOMES	National Science Education Content Standards C and D for Grades 5–8
ONE	HIV: EVOLVING MENACE	Youth develop an understanding of how a virus evolves in people.	Disease is a breakdown in structures or functions of an organism. Some diseases are the result of intrinsic failures of the system. Others are the result of damage by infection by other organisms.
			Every organism requires a set of instructions for specifying its traits. Heredity is the passage of these instructions from one generation to another.
			Hereditary information is contained in genes, located in the chromosomes of each cell. Each gene carries a single unit of information. An inherited trait of an individual can be determined by one or by many genes, and a single gene can influence more than one trait. A human cell contains many thousands of different genes.
			Millions of species of animals, plants, and microorganisms are alive today. Although different species might look dissimilar, the unity among organisms becomes apparent from an analysis of internal structures, the similarity of their chemical processes, and the evidence of common ancestry.
			Biological evolution accounts for the diversity of species developed through gradual processes over many generations. Species acquire many of their unique characteristics through biological adaptation, which involves the selection of naturally occurring variations in populations. Biological adaptations include changes in structures, behaviors, or physiology that enhance survival and reproductive success in a particular environment.
TWO	DIATOMS: ONE-CELLED WONDERS	Youth develop an understanding of the evolution of a new species in the fossil record.	The characteristics of an organism can be described in terms of a combination of traits. Some traits are inherited and others result from interactions with the environment.
			The number of organisms an ecosystem can support depends on the resources available and abiotic factors, such as quantity of light and water, range of temperatures, and soil composition. Given adequate biotic and abiotic resources and no disease or predators, populations (including humans) increase at rapid rates. Lack of resources and other factors, such as predation and climate, limit the growth of populations in specific niches in the ecosystem.
			Millions of species of animals, plants, and microorganisms are alive today. Although different species might look dissimilar, the unity among organisms becomes apparent from an analysis of internal structures, the similarity of their chemical processes, and the evidence of common ancestry.
			Biological evolution accounts for the diversity of species developed through gradual processes over many generations. Species acquire many of their unique characteristics through biological adaptation, which involves the selection of naturally occurring variations in populations. Biological adaptations include changes in structures, behaviors, or physiology that enhance survival and reproductive success in a particular environment.
			Fossils provide important evidence of how life and environmental conditions have changed.

THREE	ANTS & CO.: TINY FARMS	Youth develop an understanding of how the evolution of one species can affect the evolution of another.	An organism's behavior evolves through adaptation to its environment. How a species moves, obtains food, reproduces, and responds to danger are based in the species' evolutionary history. The number of organisms an ecosystem can support depends on the resources available and abiotic factors, such as quantity of light and water, range of temperatures, and soil composition. Given adequate biotic and abiotic resources and no disease or predators, populations (including humans) increase at rapid rates. Lack of resources and other factors, such as predation and climate, limit the growth of populations in specific niches in the ecosystem. Millions of species of animals, plants, and microorganisms are alive today. Although different species might look dissimilar, the unity among organisms becomes apparent from an analysis of internal structures, the similarity of their chemical processes, and the evidence of common ancestry. Biological evolution accounts for the diversity of species developed through gradual processes over many generations. Species acquire many of their unique characteristics through biological adaptation, which involves the selection of naturally occurring variations in populations. Biological adaptations include changes in structures, behaviors, or physiology that enhance survival and reproductive success in a particular environment.
FOUR	HAWAIIAN FLIES: SONG & DANCE SUCCESS	Youth develop an understanding of how biological evolution, acting over generations, can account for the diversity of species.	The characteristics of an organism can be described in terms of a combination of traits. Some traits are inherited and others result from interactions with the environment. An organism's behavior evolves through adaptation to its environment. How a species moves, obtains food, reproduces, and responds to danger are based in the species' evolutionary history. A population consists of all individuals of a species that occur together at a given place and time. All populations living together and the physical factors with which they interact compose an ecosystem. Millions of species of animals, plants, and microorganisms are alive today. Although different species might look dissimilar, the unity among organisms becomes apparent from an analysis of internal structures, the similarity of their chemical processes, and the evidence of common ancestry. Biological evolution accounts for the diversity of species developed through gradual processes over many generations. Species acquire many of their unique characteristics through biological adaptation, which involves the selection of naturally occurring variations in populations. Biological adaptations include changes in structures, behaviors, or physiology that enhance survival and reproductive success in a particular environment.
FIVE	GALÁPAGOS FINCHES: FAMOUS BEAKS	Youth develop an understanding of how the environment influences the evolution of species.	The characteristics of an organism can be described in terms of a combination of traits. Some traits are inherited and others result from interactions with the environment. A population consists of all individuals of a species that occur together at a given place and time. All populations living together and the physical factors with which they interact compose an ecosystem. Millions of species of animals, plants, and microorganisms are alive today. Although different species might look dissimilar, the unity among organisms becomes apparent from an analysis of internal structures, the similarity of their chemical processes, and the evidence of common ancestry. Biological evolution accounts for the diversity of species developed through gradual processes over many generations. Species acquire many of their unique characteristics through biological adaptation, which involves the selection of naturally occurring variations in populations. Biological adaptations include changes in structures, behaviors, or physiology that enhance survival and reproductive success in a particular environment.

SIX	HUMANS & CHIMPS: ALL IN THE FAMILY	Youth develop an understanding of the evolutionary relationship between humans and apes.	Every organism requires a set of instructions for specifying its traits. Heredity is the passage of these instructions from one generation to another. Millions of species of animals, plants, and microorganisms are alive today. Although different species might look dissimilar, the unity among organisms becomes apparent from an analysis of internal structures, the similarity of their chemical processes, and the evidence of common ancestry. Hereditary information is contained in genes, located in the chromosomes of each cell. Each gene carries a single unit of information. An inherited trait of an individual can be determined by one or by many genes, and a single gene can influence more than one trait. A human cell contains many thousands of different genes. Biological evolution accounts for the diversity of species developed through gradual processes over many generations. Species acquire many of their unique characteristics through biological adaptation, which involves the selection of naturally occurring variations in populations. Biological adaptations include changes in structures, behaviors, or physiology that enhance survival and reproductive success in a particular environment. The characteristics of an organism can be described in terms of a combination of traits. Some traits are inherited and others result from interactions with the environment.
SEVEN	WHALES: WALKING INTO THE PAST	Youth develop an understanding of how fossils can provide evidence for evolutionary relationships.	An organism's behavior evolves through adaptation to its environment. How a species moves, obtains food, reproduces, and responds to danger are based in the species' evolutionary history. Millions of species of animals, plants, and microorganisms are alive today. Although different species might look dissimilar, the unity among organisms becomes apparent from an analysis of internal structures, the similarity of their chemical processes, and the evidence of common ancestry. Biological evolution accounts for the diversity of species developed through gradual processes over many generations. Species acquire many of their unique characteristics through biological adaptation, which involves the selection of naturally occurring variations in populations. Biological adaptations include changes in structures, behaviors, or physiology that enhance survival and reproductive success in a particular environment. Extinction of a species occurs when the environment changes and the adaptive characteristics of a species are insufficient to allow its survival. Fossils indicate that many organisms that lived long ago are extinct. Extinction of species is common; most of the species that have lived on the earth no longer exist. Fossils provide important evidence of how life and environmental conditions have changed.

Assessment

Two kinds of assessment questions are provided in each activity. At the end of each part, a question called "Consider This" asks youth to summarize their understanding of the main point for that section. At the end of each activity, an overall assessment question, "Be a Science Reporter," asks youth to summarize their understanding of the main points of the entire activity. These questions are based on assessment guidelines provided by the National Science Education Standards (NRC 1996). Table 2 relates the learning outcomes for each activity to the "Be a Science Reporter" assessment questions.

Table 2: Overview of the Learning Outcomes and Assessment Questions for Each of the Seven Activities.

ACTIVITY NUMBER	ACTIVITY TITLE	LEARNING OUTCOMES	ASSESSMENT: Be a Science Reporter
ONE	HIV: EVOLVING MENACE	Youth develop an understanding of how a virus evolves in people.	Write a short news story about HIV. Tell your readers about how a baby who is born with HIV carries exactly the same strains of the virus as the mother. Within six months, the baby can have millions of new strains. Based on what you have learned, explain how you think the baby gets new strains of HIV.
TWO	DIATOMS: ONE-CELLED WONDERS	Youth develop an understanding of how a new species emerges.	Write a short news story about diatoms. Tell your readers about a new species of diatom that is found only in Yellowstone Lake. From core samples taken beneath the lake, scientists can tell that the Yellowstone diatom first appeared thousands of years ago, at a time when the climate was dramatically changing. Based on what you have learned, explain how you think this new diatom came to exist in the lake.
THREE	ANTS & CO.: TINY FARMS	Youth develop an understanding of how organisms evolve to be dependent on one another.	Write a short news story about ants and their partners. Tell your readers about how leaf-cutter ants, their fungus crop, the crop pest, and the bacteria have lived together in an association for millions of years. Based on what you have learned, explain how you think this association came about. What evidence supports your explanation?
FOUR	HAWAIIAN FLIES: SONG & DANCE SUCCESS	Youth develop an understanding of how evolution accounts for the diversity of species over many generations.	Write a short news story about Hawaiian flies. Tell your readers about how eight million years ago, there were no Drosophila flies on Hawaii. Now there are more than 800 species found only on the islands. Based on what you have learned, explain how you think so many new species came to be on the islands.
FIVE	GALÁPAGOS FINCHES: FAMOUS BEAKS	Youth develop an understanding of how the environment influences the evolution of species	Write a short news story about the medium ground finches on the island of Daphne Major. Tell your readers about how the drought of 1977 led to changes in the characteristics of the finch population there. Based on what you have learned, explain why you think after the drought the finch population in the next generation had larger beaks.
SIX	HUMANS & CHIMPS: ALL IN THE FAMILY	Youth develop an understanding of the evolutionary relationship between humans and apes.	Write a short news story about humans and chimpanzees. Tell your readers about how new DNA studies from humans and chimpanzees suggest they are close relatives. Based on what you have learned, explain how you think a modern chimpanzee and a modern human could have a common ancestor.
SEVEN	WHALES: WALKING INTO THE PAST	Youth develop an understanding of how fossils can provide evidence for evolutionary relationships.	Write a short news story about whales. Tell your readers about new fossil discoveries from Pakistan that suggest whales are related to animals that once lived on land. Based on what you have learned, explain how you think a whale could have an ancestor that lived on land.

Table 3: Overview of the Materials Required for Each of the Seven Activities.

ACTIVITY NUMBER	ACTIVITY TITLE	MATERIALS PART ONE	MATERIALS PART TWO	MATERIALS PART THREE
ONE	HIV: EVOLVING MENACE	Each team of 2 will need: Virus Flip book sheets 1, 2, and 3, clip or stapler, marker, colored pencils, scissors.	Each team of 2 will need: Virus Mutation Tracker sheet, Available "vaccine," Your "vaccine," coin, colored pencils: yellow, red, blue, green.	Each team of 2 will need: HIV Fact Cards, poster-sized paper, collage materials or magazines, glue stick, markers or colored pencils, scissors.
TWO	DIATOMS: ONE-CELLED WONDERS	Each team of 2 or more will need: Diatom Fact Cards 1–3, sheet of paper, tape, scissors.	Each team of 2 will need: Your Pollen Chart, Your Diatom Chart, Read the Core Sample, one small cup (5 oz, clear is best), one clear plastic straw, brown sugar (9 teaspoons), three different bright colors of sugar (1/2 teaspoon each), paper plates (3), plastic spoon, cm ruler, newspapers.	Each team of 2 will need: Core Sample Chart, Pollen Chart (from Part Two).
THREE	ANTS & CO.: TINY FARMS	Each team of 2 will need: Ant Fact Cards 1 and 2, tape or glue stick, large sheet paper (about 11x17"), scissors.	Each team of 2 will need: Crop Pest vs. Bacteria Experiment, Crop Pest vs. Bacteria Data Chart, cm ruler, 2 colored pencils (red plus one other).	Each team of 2 will need: Crop Pest vs. Bacteria vs. New Pest Experiment, Crop Pest vs. Bacteria vs. New Pest Data Chart, cm ruler, 3 different colored pencils (red plus two others).
FOUR	HAWAIIAN FLIES: SONG & DANCE SUCCESS	Each team of 2 will need: Fly Fact Cards 1–3, paper, tape or glue, scissors.	Each player will need: song simulators, Songs of Hawaiian Drosophila card (either a male or a female), chair, scissors.	Each team of 2 will need: Fly paper, Tracking Fly Population Chart, scissors.
FIVE	GALÁPAGOS FINCHES: FAMOUS BEAKS	Each team of 2 will need: 6 Finch Beak Sheets, Finch Beak Measurements chart, compass, ruler (cm and mm).	Each team of 2 will need: Battle of the Beaks: Normal Year Chart, Battle of the Beaks: Drought Year Chart, 4 popsicle sticks, 2 rubber bands, 9 pennies, 1 teaspoon large seeds (garbanzos), 1 teaspoon small seeds (mustard seeds), cm ruler, tape, sheet of paper, paper plate, timer.	Each team of 2 will need: Seed Abundance Graph, Finch Population Graph.
SIX	HUMANS & CHIMPS: ALL IN THE FAMILY	Each team of 2 or more will need: Chimpanzee vs. Human sheet, Comparing Creature Features sheet, Feature Cards (blanks; one sheet per group), Hidden Feature Cards (one sheet per group), tape, scissors.	Each team of two will need: Generations of Copies sheet, scissors.	Each team of two will need: Chimp vs. Human DNA Sequences sheets 1 and 2 (taped together), scissors.
SEVEN	WHALES: WALKING INTO THE PAST	Each team of 2 will need: Pretzel Species Sheet, stiff paper or cardboard, liquid glue and EITHER: 1 pretzel broken into 3–4 pieces put inside an envelope OR cookie dough baked with 3–4 pretzel pieces & a big nail or nail file.	Each team of 2 will need: One Mystery Fossil Bones card (cut separate cards for Fore Limbs, Hind Limbs, Skull, OR Teeth, OR Neck, and Ribs); one matching Some Known Bones sheet (Fore Limbs, Hind Limbs, Skull, and Teeth, OR Spine, Neck, and Ribs); one Sum It Up sheet (Fore Limbs and Hind Limbs OR Skulls and Spine), scissors.	Each team of 2 will need: Creature Features: Comparing Anatomy sheet.

Materials

Each activity uses readily available materials. Preparation time for the activities is under 20 minutes. It is best if each user has a copy of the activity book to read and write in. But a group can also share a book if the "consumable" pages—those that require written answers or sections cut out—are copied in advance. These pages are listed at the beginning of each materials list. Masters for the activities are available at no cost from the NSTA and Explore Evolution websites (*www.nsta.org* or *http://explore-evolution.unl.edu*). A pencil or pen is required for all activities and is not listed on the materials list.

Chapter 1

Making Sense of Evolution

Carl Zimmer

At first glance, a whale doesn't seem to have anything in common with a fly. But whales and flies are both living things, and that means they actually share a great deal in common. They use the same sorts of molecules to generate energy and keep themselves alive. They also share a common history, belonging to different branches of the same tree of life—a tree that is the product of some four billion years of evolution.

This book is intended to show how scientists study evolution in the different branches of the tree of life. While the details that they study are different from species to species, evolutionary biologists seek to understand how the same fundamental principles of evolution have produced the stunning diversity of life that surrounds us today. In the present chapter, we'll take a look at the basic principles by which evolution works, and consider how they have interacted to produce the organisms you will encounter in the activities in this book.

DNA

At the heart of evolution—at least on this planet—is a string-shaped molecule called deoxyribonucleic acid. You probably know it by its abbreviation: DNA. DNA has the extraordinary ability to store information about building and maintaining a living organism. More extraordinarily, it can also be copied, making it possible for an organism to reproduce itself. And even more extraordinarily still, the information encoded in DNA can change. Over millions of years, the DNA passed down through a lineage can undergo dramatic changes. This type of change makes evolution possible

DNA has the shape of a double helix. Two long chains of compounds twist around each other. These "backbones" are linked at regular intervals by pairs of compounds called nucleotides. These nucleotides store the information in DNA, just as letters store the information in a cookbook. Each nucleotide can have one of four alternative forms: adenine, cytosine, guanine, and thymine. (Biologists tend to abbreviate their names down to A, C, G, and T.) Thanks to their molecular structure, each

A guide to DNA

DNA is the molecule that stores the recipe needed to produce an organism.

DNA strands
DNA is made up of two backbones and their bases, which are known as **strands**. These strands can be pulled apart with high heat or harsh chemicals.

Backbone is made of sugars and phosphates. It forms the outside edges of the DNA molecule.

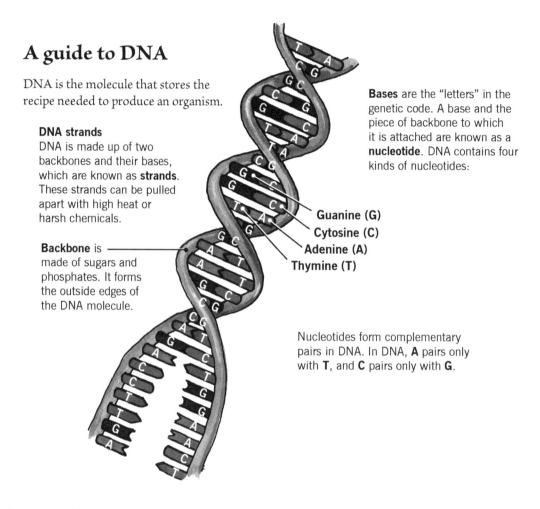

Bases are the "letters" in the genetic code. A base and the piece of backbone to which it is attached are known as a **nucleotide**. DNA contains four kinds of nucleotides:

Guanine (G)
Cytosine (C)
Adenine (A)
Thymine (T)

Nucleotides form complementary pairs in DNA. In DNA, **A** pairs only with **T**, and **C** pairs only with **G**.

Science Museum of Minnesota (SMM) Adam Wiens, Lonnie Broden illustration.

nucleotide can only attach to one other nucleotide. A only attaches to T, and C to G.

An entire molecule of human DNA contains over 3 billion nucleotides. If a publisher printed the code in regular hardcover books, it could fill a set of 1,000 volumes, each 1,000 pages long. To the untrained eye, this sequence would look like an epic of gibberish. But our cells can decode DNA and use it like a cookbook to build the molecules called proteins that make up our bodies.

The human body contains about 100,000 different kinds of proteins. Each one is composed of a string of compounds known as amino acids. Unlike DNA, which uses four different nucleotides, proteins can be built from twenty different amino acids. Some of the amino acids in a protein are attracted to one another, while others repel one another. This attraction and repulsion causes a protein to fold and twist in on itself, taking on an intricate structure. The structures of proteins allow them to do all sorts of jobs in the body. Some proteins in the stomach help break down the food we eat. Other proteins can clutch oxygen molecules and carry them through the blood. Other proteins give the body structure, producing hair, fingernails, and skin. Others carry signals from one cell to another, so that each cell can respond to changes in the environment.

DNA contains the instructions for building proteins. Each protein is encoded in a specific stretch of DNA, known as a gene. The first step in building a protein is to copy the information in a gene. Special DNA-reading proteins grab onto the gene and produce a new molecule with the same sequence of nucleotides as the gene. Unlike the double-stranded structure of DNA, this copy is a single-stranded version called RNA, which is short for ribonucleic acid. As the DNA-reading proteins move along the gene, they skip over many parts of its sequence. As a result, the RNA molecule contains only an edited version of the gene's DNA sequence.

This RNA molecule is then ferried to a protein-building factory in the cell called a ribosome. It may be hard to imagine how the ribosome can use instructions written in the four-letter language of genes to build proteins, which are made out of twenty different amino acids. Our ribosomes manage this feat by reading RNA three nucleotides at a time. Different triples of nucleotides cause the ribosomes to produce different amino acids. AGG codes for alanine, for example, while GGG codes for glycine.

It has been just over fifty years since scientists discovered the structure of DNA, and in that time, they've made stunning progress in understanding how this molecule makes life possible. But many mysteries remain. At last count, researchers estimate that humans carry a total of 20,000 to 25,000 genes. How then do our bodies produce 100,000 proteins? It turns out that our cells can use a single gene to make many different proteins, simply by editing the RNA in different ways. Exactly how a cell "decides" to edit its genetic code is a question that will keep molecular biologists busy for years to come.

Despite these open questions, it is now clear that DNA plays a crucial role in evolution. When genes mutate, the proteins they produce may change structure. As a result, the protein may take on a different function. If the new protein helps an organism survive longer and produce more offspring, the mutated gene may become more common over many generations. Over millions of years the rise and spread of new genes produces evolutionary change. We'll explore this process in more detail in the rest of this chapter.

Heredity

We are all mortal, but the information in our DNA is potentially immortal. That's because it is inherited in our children, who can pass it on to our grandchildren, and so on, ad infinitum.

The details of heredity vary from species to species, but fundamentally they're the same in all life. The simplest place to start is with bacteria. These single-celled microbes have DNA, as we do, and they use it to produce proteins as we do. Bacteria replicate by dividing in two, and during their preparations for division, they make a copy of their DNA. Special proteins pull apart the two DNA strands, separating the nucleotides in the process. To each of these split DNA molecules, the proteins add the corresponding nucleotides. When they are finished, they have produced two nearly identical sets of DNA. (We'll get back to "nearly" in a little while.) Once the two sets of DNA are assembled, the bacterium stretches until it

Making protein from DNA

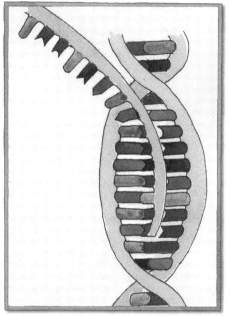

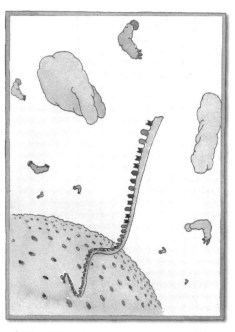

1. Special proteins can read the nucleotides of DNA that make up a gene. They make a temporary copy of the gene known as RNA.

2. The RNA gets ferried to a protein-building factory

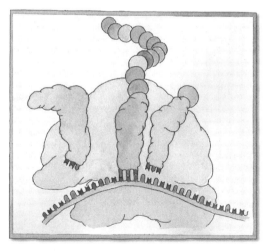

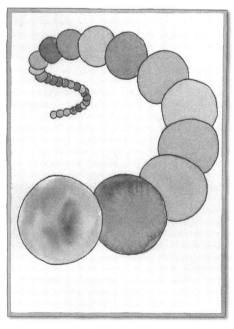

3. There, the RNA guides the assembly of a new protein.

4. The new protein.

SMM Adam Wiens, Lonnie Broden illustration.

pulls apart into two separate microbes, each of which contains DNA. Both of the resulting microbes use the same genes to produce the same proteins as in their ancestors.

Our own cells divide in this way as well. Most of the cells in our bodies contain two copies of each gene, one version we inherit from our mother, the other from our father. When our cells divide, they copy both sets of genes, so that each daughter cell ends up with a full complement of DNA. As our bodies grow, they produce new cells this way. But all of these cells can only survive in our own bodies. When we die, they die.

A tiny fraction of the cells in our body are exceptions to this rule. They are the cells that help give rise to children—eggs in women and sperm in men. Eggs and sperm are produced by special precursor cells that divide in a peculiar way. As a precursor cell prepares to split, its two sets of DNA come together. Pairs of genes line up. Some of them trade places, and the two DNA molecules then pull apart, with their genes now shuffled between each other. The precursor cell then divides. But it does not make an extra copy of its DNA before it does so. Instead, each of the new cells that are formed inherits only a single set of genes.

During fertilization, a male sperm fuses to a female egg in a woman's body. The sperm's set of genes combines with the egg's to create a full set of DNA. The fertilized egg starts to divide to form an embryo, and each of the new cells receives the two sets of genes.

The structure of DNA thus allows it not only to encode the information needed to keep an organism alive, but also to allow an organism to create a new individual with its own set of DNA. But as we'll see below, this copying process is an imperfect one, and that imperfection is essential for evolution.

Variation

Imagine that you were handed the thousand-volume record of your genome. Imagine that the book was designed so that the sequences of each pair of genes were lined up together so that you could compare them nucleotide by nucleotide. You would be able to see that these sequences are almost—but not entirely—identical. You might have inherited a version from your mother that is slightly longer than your father's, due to a few extra nucleotides that appear in the middle. In other cases, the genes might differ at a single position. Let's say you scan two version of the same gene, position by position. In the 300th position, your father's version of a gene might contain guanine, while a cytosine might occupy that same position in your mother's gene.

Often these variations don't make much of a difference. The differences may lie in parts of the gene that are ignored by our cells as they produce proteins. But occasionally these differences are important. You carry genes that build special receptors on the surface of your cells, for example. Your mother's copy of a particular receptor gene might be normal, but your father's gene might have a peculiar sequence that causes a ribosome to stop reading it, so that it cannot produce a viable protein. As a result, the only receptors in your body are produced by your mother's gene.

If you have a child, you can only give it one version of this receptor gene. It's strictly a matter of chance whether you will pass down your father's faulty gene or your mother's working one. Now imagine that your spouse also carries a working and faulty version of the

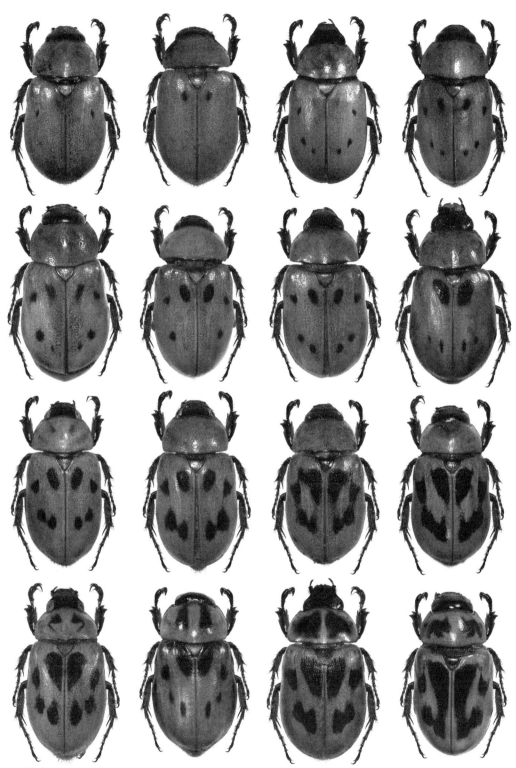

Variation within a single beetle species, *Cyclocephala sexpunctata*.

Photo courtesy Angie Fox. Used with permission from the Division of Entomology at the University of Nebraska State Museum.

Making Sense of Evolution

same gene. (We'll call the working copy A and the faulty one a.) Your spouse also has a 50-50 chance of passing down either version to your child. As a result, your child may end up with either two A's, two a's, or one A and one a. The odds of these combinations are, respectively, 25%, 25%, and 50%. If you have four children, in other words, you'd expect, on average, that one would have two working copies, one would have no working copies, and the other two children would each have only one working copy.

This same roll of the dice governs which version of every gene you inherit from your parents. As a result, two siblings will only share 50% of the same versions of all their genes. That's why siblings can look so different from one another—except, of course, for identical twins, which are produced from a single fertilized egg.

Now let's go back to those two different versions of the same gene that you inherited from your parents. Where did those differences come from?

To understand the source of these variations, it helps to turn again to bacteria, because they reproduce so simply. Bacteria only have a single copy of each of their genes. When a single bacterium divides, the two resulting bacteria each inherit their own copy of the same gene. If the genes are perfectly copied, these descendants are thus genetically identical to their ancestors.

But DNA-copying proteins are not perfect. Sometimes they accidentally switch a nucleotide. Sometimes they skip a few nucleotides. Sometimes they duplicate a segment. These changes to the genetic code are called mutations. Mutations strike randomly, but that doesn't mean they are completely unpredictable. If you flip a coin once, you can't predict whether it will land heads or tails. But if you flip 1,000 coins, you can be quite sure that it will come up close to 500 times heads and 500 times tails. Likewise, scientists can calculate the rate of mutations in bacteria and other organisms. In E. coli bacteria, for example, each nucleotide pair has a 1 in 10 trillion chance of mutating with each cell division. Those odds are tiny, but you must bear in mind that each bacterium contains five million nucleotide pairs, that bacteria can reproduce as often as three times in an hour, and that a single petri dish may contain billions of bacteria.

Humans mutate like bacteria and all other organisms. One reason we know this is that sometimes perfectly healthy parents can have a child with a genetic disorder. The disorder stems from a mutant gene carried by the child, but which neither parent carries. Many mutations are completely harmless, however, because they strike parts of the DNA that are not essential for making proteins. And some mutations are even beneficial.

The processes described in this section ensure that in every generation a population will have genetic variation. In the figure on page 6, the beetles show a wide variation in the patterns on their wings. Yet they all belong to the same species, and can thus interbreed and pass on their genes to the next generation. As we'll see later in this chapter, this variation makes evolution possible, because some variations can allow an organism to have more offspring than other variations.

Selection

A newly mutated gene may disappear with the organism that carries it. Some mutations cause devastating effects on the development of embryos, for example, killing them before they can be born. On the other hand, many mutations cause little or no change to the way our genes work. A person who acquires one of these "neutral" mutations can grow up and have children, who may, in turn, inherit the mutant gene.

Mutations that are beneficial gradually become more common in a population. What does it mean to be beneficial? The answer is both simple and hugely complex. The simple version is this: A mutation that allows an individual, on average, to reproduce more successfully than other members of its own species is beneficial. But what determines reproductive success can vary from place to place and from time to time.

Scientists can watch the spread of beneficial mutations by studying bacteria. At Michigan State University, for example, microbiologist Richard Lenski started a colony of bacteria in 1988. From a single founding bacterium, he established twelve identical colonies. The bacteria were all fed a very low supply of sugar. Any mutation that could help the bacteria survive on this meager diet would allow them to reproduce faster. Any mutation that made it harder for them to turn the sugar into energy would slow down their reproduction.

Periodically, Lenski and his colleagues took a sample from each colony and put them in a freezer. Once frozen, the bacteria go into suspended animation. Meanwhile, the other bacteria in their colony continued eating sugar, reproducing, mutating, and eating more sugar. Lenski could then take some of the ancestral bacteria out of the freezer, revive them, and watch how well they reproduced compared to later generations.

After 30,000 generations (seventeen years), the results are clear: Beneficial mutations have steadily increased the ability of the bacteria to reproduce. The bacteria that make up Lenski's colonies today can reproduce about twice as quickly as they could in 1988. This pattern has repeated itself in all twelve colonies that Lenski established from a single ancestor.

Lenski's results indicate that when an individual bacterium acquires a beneficial mutation, its descendants outcompete the rest of the colony that lack the mutation. Over time, the mutants take over the entire colony, as the other lineages eventually die out. Then a new beneficial mutation strikes one of the descendants, making it even better at reproducing, and the cycle repeats itself.

Even in a simple bacterial colony, however, it takes a lot of work to decipher the spread of beneficial mutations. Over the length of the experiment, literally billions of mutations have arisen in each colony. But Lenski estimates that only 50 of those mutations were beneficial enough to have spread through the entire population and survive until today. Figuring out what benefit each of those 50 mutations provides will take many years of research. But the fact remains that beneficial mutations have profoundly transformed these bacteria in a matter of years.

Obviously, this sort of study is impossible in humans. We take so long to reach sexual maturity that 30,000 human generations represents more than half a million years. To study the spread of beneficial mutations in humans, scientists have to look for indirect evidence.

One group of beneficial mutations appears to have spread over the past few thousand years as a result of the domestication of cattle. Cattle herding provided people with an immediate benefit, in the form of a reliable supply of meat. Cows also produce milk, which is also a good source of nutrition, though the first cattle herders probably couldn't drink much of it. As infants, most mammals digest their mothers' milk with the help of a protein they produce, called lactase. Lactase acts like a pair of scissors, snipping apart a sugar called lactose found in milk. The snipped pieces of lactose can then be absorbed into the mammal's bloodstream. But as mammals grow older, they stop producing lactase and can no longer digest milk. Unable to drink milk, they shift to their adult diet. Our distant ancestors were also unable to drink milk as adults, and even today many ethnic groups cannot. Their lactose intolerance gives them indigestion when they try to drink milk or eat cheese.

People who descend from traditional cattle herders, on the other hand, are much more likely to be lactose tolerant. Geneticists have identified a short sequence of DNA called LCT that has a different sequence in people who are lactose tolerant and intolerant. Scientists suspect that the new mutations disabled the off-switch for lactase production. They allowed people to continue drinking milk into adulthood, and the extra nutrition ultimately translated into extra children. As a result, this beneficial mutation spread through these cattle-herding populations.

When Charles Darwin first began to think about evolution, he couldn't know about DNA, mutations, and all the other insights of molecular biology. All he could see were the outward effects of mutations and genetic variation. In every generation of humans or any other species, individuals showed variation in their traits. If individuals with a particular trait reproduced more than other members of a population, then their trait should become more common over time. Humans had already demonstrated that this change was possible in

the way they bred animals and plants. A pigeon fancier who wanted to raise a particular breed of pigeon would only produce eggs from those birds with the longest feathers or the brightest plumage or whatever trait was admired in the breed. The differences between a drab wild pigeon and its extravagant domesticated cousins were obvious.

Darwin argued that just as humans carry out artificial selection, wild animals and plants undergo natural selection. Traits that helped a wild individual survive and reproduce would become more common over time. And with enough time, evolution by natural selection could produce all of life's diversity.

Time

Darwin believed that natural selection acted too slowly to be noticed in our ordinary lives. It had taken vast periods of time for a common ancestor to evolve into species as diverse as a human and a butterfly. But Darwin was confident that the Earth was millions, if not billions of years old. In the mid-1800s this was the emerging consensus among geologists, but there was still a lot of room for doubt. The great physicist Lord Kelvin argued that the Earth must have cooled from a ball of molten rock, and that if the world was billions of years old it should have cooled to a much lower temperature than could be measured in deep mine shafts. Instead, he concluded that the Earth was less than 20 million years old.

It wasn't until the 20th century that physicists discovered a precise way to estimate the age of the Earth and its fossils. Certain elements, such as uranium, are unstable and break down into other elements. These so-called radioactive atoms break down at predictable rates over millions or billions of years. Their clock-like rate of decay allows scientists to measure the ages of rocks that contain them.

Thanks to radioactive clocks, scientists have determined that the Earth is indeed billions of years old—4.55 billion years old, to be exact. The oldest generally accepted evidence of life on Earth dates back 3.7 billion years. This evidence does not take the form of a fossil, however. Carbon atoms come in light and heavy forms (the heavy form contains extra neutrons). Plants and bacteria that absorb carbon dioxide from the air prefer light carbon to heavy, and after they die, their remains can become incorporated into rocks. These rocks retain a distinctive ratio of light to heavy carbon atoms. And the oldest rocks that have this ratio can be found in Greenland, dating back 3.7 billion years. (The organisms that produced this ratio were almost certainly photosynthetic bacteria. The oldest plant fossils are only 475 million years old. Fossils of microbes date back at least 3.4 billion years.)

The oldest fossils of animals—jellyfish-like creatures—are about 575 million years old. Our own ancestors moved from water to land about 360 million years ago. Since then, land vertebrates have diversified into amphibians, reptiles, birds, and mammals. Primates, the order of mammals to which human belong, are known from fossils dating back 55 million years. The living primates most closely related to humans are the great apes—chimpanzees, bonobos, gorillas, and orangutans. Fossils of early apes suggest that this common ancestor lived about 13 million years ago. Chimpanzees, bonobos, and humans share a much more recent common ancestor that lived an estimated 5 to 6 million years ago. (We'll describe the methods for making that estimation later in this chapter.) After our ancestors split from chimpanzees, they

gave rise to as many as twenty human-like species in addition to our own. The oldest fossils that clearly belong to *Homo sapiens*—our own species—are 195,000 years old.

It is now clear that the world is vastly old, as Darwin had suspected. Mutations, variations, and natural selection have been interacting with one another for billions of years. But scientists have found that evolution has actually played out at many different time scales. In some cases, scientists can even document effects of natural selection in a matter of years or even days.

Evidence for Evolution

Scientists have amassed a vast amount of evidence that life has evolved over the past 3.7 billion years, and that our own species is a product of that evolution along with the estimated 5 to 100 million other species with which we share the planet. The theory of evolution is so well established that it has become part of the bedrock on which modern biology is built. You can see its importance simply by skimming the leading journals in a wide range of life sciences. In a medical journal you may encounter an evolutionary tree showing how strains of a hepatitis virus have evolved from a common ancestor. In a microbiology journal you might read about how a species of bacteria has adapted to feeding on human pollution over the past 50 years. A journal dedicated to RNA may include a paper presenting evidence suggesting that some versions of RNA are relics of the earliest forms of life that lived 3.7 billion years ago. An ecology journal may examine how a weed that was accidentally introduced to the United States has adapted to its new home.

Some people may be puzzled when scientists accept the reality of evolution, given that neither they nor the scientists were around billions of years ago to watch it take place. It's important to remember that scientists are not always eyewitnesses to the phenomena they seek to explain. Physicists, for example, do not have to actually see an electron with their own eyes to be confident that it exists. Instead, they look for the effects of electrons through experiments. Geologists cannot see lava rise from miles below the Earth's surface before it erupts out of a volcano's crater. But they can infer the history of an eruption by looking at many lines of evidence, from the chemistry of the lava itself to recordings of seismic activity around the volcano. Scientists study evolution in the same way, by looking at the evidence that's available, whether it is a fossil in the ground or bacteria in a lab.

Some people may also be puzzled by the way scientists refer to the "theory of evolution." If it's just a theory, how can scientists accept it so readily? Scientists don't use the word *theory* in the informal way most people do, meaning a guess or a hunch. A scientific theory is an overarching explanation for some aspect of the natural world (one that has been verified many times by different groups of scientists) that makes sense of what would otherwise be a mysterious collection of data. Scientific theories give rise to hypotheses, which can be tested through experiments or observations. The more a scientific theory holds up to this sort of scrutiny, the more it becomes accepted.

Modern science is dominated by theories, from the theory of quantum mechanics to the germ theory of disease to the theory of evolution by natural selection. All of these theories have withstood serious scientific scrutiny and have come to be embraced by the scientific community. Scientists did not have to wait until every single ramification of these theories had been explored before accepting them.

Every major scientific theory still leaves many mysteries for scientists to decipher, but that does not take away from their importance.

Over the past 150 years, scientists have gathered evidence for evolution from different fields, ranging from paleontology to genetics. I will consider here just one example of this evidence: the tree of life.

Charles Darwin claimed that new species budded off from older ones like the branches on a tree. Many of the similarities seen in a group of animals, such as mammals, were due to a common ancestry.

Consider a group of land vertebrates: snakes, lizards, crocodiles, birds, whales, camels, humans, and chimpanzees. Scientists have compared their physical features--their skeletons, muscles, organs, cell types, and so on--and drawn an evolutionary tree. Humans and chimpanzees share a close common ancestor, reflected in unique traits they share in their brain structure, cell receptors, and so on. Humans and chimpanzees are more distantly related to other mammals--camels and whales, for example. All of these mammals share mammary glands to produce milk and a placenta to support embryos. Birds, snakes, crocodiles, and lizards lack this unique combination of traits. They all share a common ancestor of their own. But if you move far back enough on the tree, the branches of all of these land vertebrates come together. As different as these animals may be, they still share certain traits, such as three distinctive membranes that form around their embryos that aren't found on frogs, salamanders, or fish.

When scientists began to obtain DNA sequences from these animals in the 1980s and 1990s, they began to compare the sequences to see what they had to say about the evolution of land vertebrates. These studies produced an almost identical tree to the one scientists had drawn based on physical characteristics. To appreciate just how important this discovery was, you have to bear in mind all of the possible ways such a tree could have been drawn. It might have shown whales more closely related to humans than to camels, for example, or perhaps sharing a close ancestry with birds. There are literally millions of possible trees that scientists could have derived by studying the DNA of these animals. But of all these possibilities, the tree they got showed the same overall pattern found in studies on physical traits. It's a beautiful example of how science works: Scientists use evolutionary theory to come up with a hypothesis, and then test it with a different set of evidence.

Origin of the Universe

Astronomers estimate that the universe is 14 billion years old. Stars and galaxies formed within the first billion years, producing carbon and other elements essential for life. Our own solar system took shape 4.5 billion years ago, setting the stage for evolution on Earth.

Whales

About 55 million years ago, the ancestors of modern whales were four-legged land mammals. Over the following 10 million years, they evolved into legless, ocean-going creatures.

Drosophila heteroneura
Hawaiian fly

Hawaiian Flies

Drosophila flies arrived on the Hawaiian Islands several million years ago, not long after the islands began to form. Since then, the flies have evolved into more than 800 species that occur nowhere else on Earth.

Rodhocetus
47.5-million-year-old whale

Millions of years

Diatoms

Diatoms are one-celled algae. Scientists have found a new species of diatom in Yellowstone Lake. They estimate that it evolved from another diatom species between 14,000 and 10,000 years ago.

HIV

HIV

Human immunodeficiency virus (HIV, the cause of AIDS) can pass from an infected mother to her child. HIV can evolve into new forms in the child in a single day.

Stephanodiscus yellowstonensis
Yellowstone diatom

Thousands of years

Days

Billions of years

Sulfolobus solfataricus
Microbe belonging to the Archaea

Early Life

The oldest evidence of life on Earth dates back 3.7 billion years. Since then, life has evolved into its astonishing variety, from microbes to dinosaurs. Scientists estimate that there are at least five million species of living organisms today and that, over the history of the Earth, perhaps 100 million species have existed.

Ants and Fungus

About 45 to 65 million years ago, ants raised a fungus as food and the fungus became dependent on the ants for survival. This evolutionary partnership is now known to include two other organisms.

Farming ants and their fungus crop

Human and chimpanzee

Humans

Humans shared a common ancestor with chimpanzees and bonobos about five to six million years ago. Since then, the two species have become very different. Humans have evolved large brains, language, and other unique features.

Years

Finches

In 1977, finches on the Galápagos Islands were hit by a harsh drought. The drought killed plants that produced the small seeds that made up much of the finches' diet. Finches with larger beaks survived more easily because they could crack open and eat the tough seeds of plants that the drought didn't affect. As a result, over the next few years the average beak size of Galápagos finches increased.

Medium ground finch from
the Galápagos Islands

SMM illustration. Archaeobacteria photo courtesy Paul Blum, *Rodhocetus* John Klausmeyer illustration, Virus, Diatom, Ant, Flies, Finch and Human/Chimp. UNSM Angie Fox illustration.

The Genealogy of a Killer

HIV causes AIDS, one of the greatest threats to public health worldwide today.
The disease evolved from viruses that infect African primates.

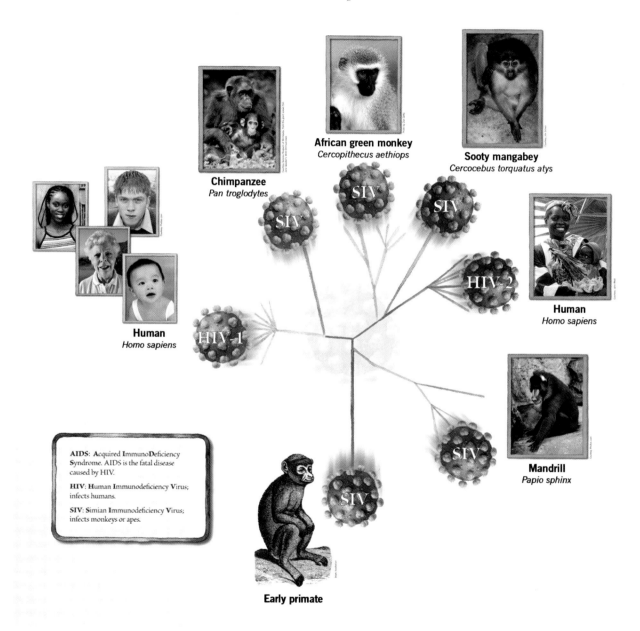

Chimpanzee
Pan troglodytes

African green monkey
Cercopithecus aethiops

Sooty mangabey
Cercocebus torquatus atys

Human
Homo sapiens

Human
Homo sapiens

SIV

SIV

SIV

HIV-2

HIV-1

SIV

Mandrill
Papio sphinx

AIDS: Acquired ImmunoDeficiency
Syndrome. AIDS is the fatal disease
caused by HIV.

HIV: Human Immunodeficiency Virus;
infects humans.

SIV: Simian Immunodeficiency Virus;
infects monkeys or apes.

SIV

Early primate

Timeline of Diatom Evolution

Deep beneath the surface of Yellowstone Lake, scientists have found the most complete fossil record of the evolution of a new species.

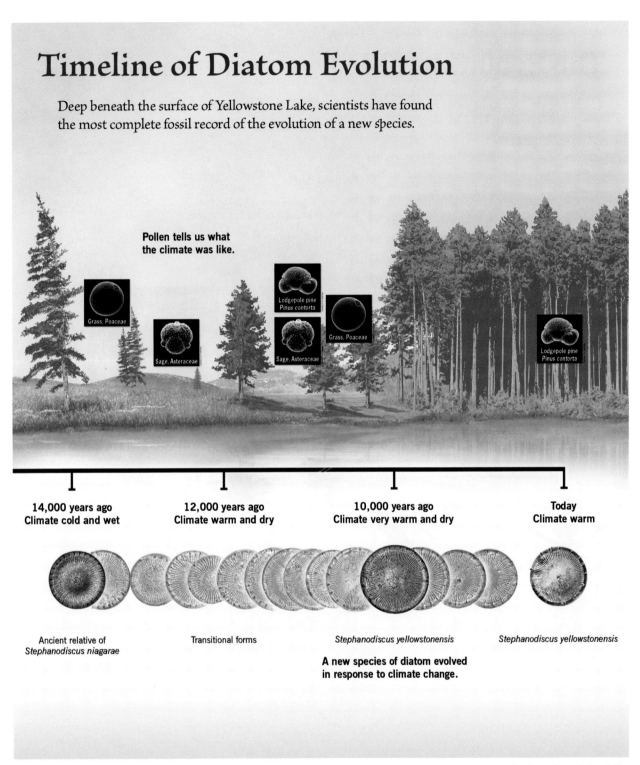

Pollen tells us what the climate was like.

Grass, Poaceae

Sage, Asteraceae

Lodgepole pine *Pinus contorta*

Sage, Asteraceae

Grass, Poaceae

Lodgepole pine *Pinus contorta*

14,000 years ago Climate cold and wet

12,000 years ago Climate warm and dry

10,000 years ago Climate very warm and dry

Today Climate warm

Ancient relative of *Stephanodiscus niagarae*

Transitional forms

Stephanodiscus yellowstonensis

Stephanodiscus yellowstonensis

A new species of diatom evolved in response to climate change.

UNSM Angie Fox and SMM illustration, *S. yellowstonensis* and Transitional Diatom photos courtesy Edward Theriot, *S. niagarae* photo courtesy Jeffery Robert Stone, Lodgepole Pine Forest photo courtesy Chris Schnepf, University of Idaho. www.forestryimages.org. Modified with permission by UNSM Angie Fox. Pollen photos courtesy Dr. Heidemarie Halbritter

The Evolution of Friends and Enemies

Biologist Cameron Currie has discovered how certain species of ants evolve in partnership with the fungi they farm, with pests that attack their crops, and with bacteria that defend against the pests.

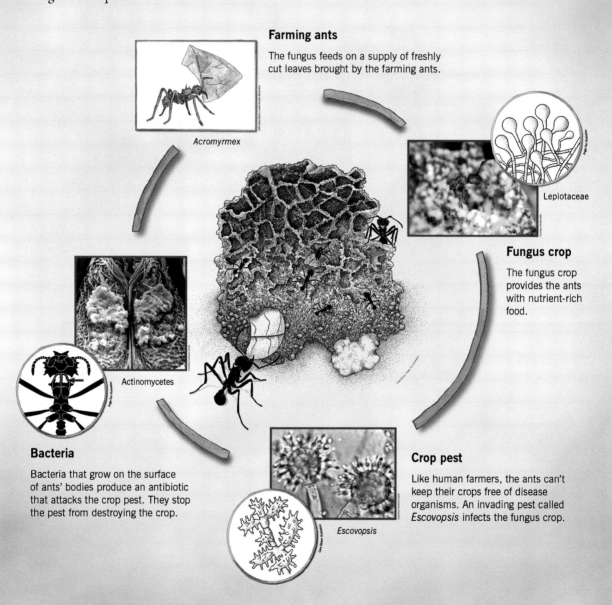

Farming ants

The fungus feeds on a supply of freshly cut leaves brought by the farming ants.

Acromyrmex

Lepiotaceae

Fungus crop

The fungus crop provides the ants with nutrient-rich food.

Actinomycetes

Bacteria

Bacteria that grow on the surface of ants' bodies produce an antibiotic that attacks the crop pest. They stop the pest from destroying the crop.

Escovopsis

Crop pest

Like human farmers, the ants can't keep their crops free of disease organisms. An invading pest called *Escovopsis* infects the fungus crop.

UNSM Angie Fox and SMM illustration. Adapted with permission from Cameron Currie/Cara Gibson. Ant on nest Cara Gibson illustration, modified by Angie Fox; Ant with leaf SMM Adam Wiens illustration; Fungus crop & Bacteria Angie Fox illustrations; Ant on Fungus crop courtesy of Ainslee Little, Crop pest and Bacteria photos courtesy of Cameron Currie; Crop pest Cara Gibson illustration.

Hawaii as a Nursery of Evolution

More than 800 species of *Drosophila* flies live on the islands of Hawaii and nowhere else on Earth.
They may all descend from a single pregnant fly that came there several million years ago.

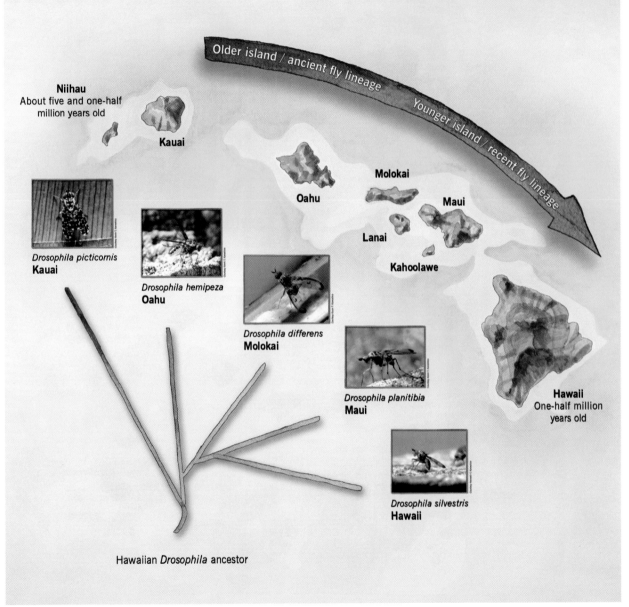

Older island / ancient fly lineage

Younger island / recent fly lineage

Niihau
About five and one-half
million years old

Kauai

Oahu

Molokai

Maui

Lanai

Kahoolawe

Drosophila picticornis
Kauai

Drosophila hemipeza
Oahu

Drosophila differens
Molokai

Drosophila planitibia
Maui

Hawaii
One-half million
years old

Drosophila silvestris
Hawaii

Hawaiian *Drosophila* ancestor

SMM Lonnie Broden illustration, Adapted from FREEMAN, SCOTT; HERRON, JON, EVOLUTIONARY ANALYSIS, 3rd Edition, © 2004, Reprinted by permission of Pearson Education, Inc.,
Upper Saddle River, NJ. *Drosophila picticornis, D. hemipeza, D. differens, D. planitibia* photos courtesy Kevin T. Kaneshiro, *Drosophila silvestris* photo courtesy Kenneth Y. Kaneshiro.

Finch Diversity: A Result of Evolution

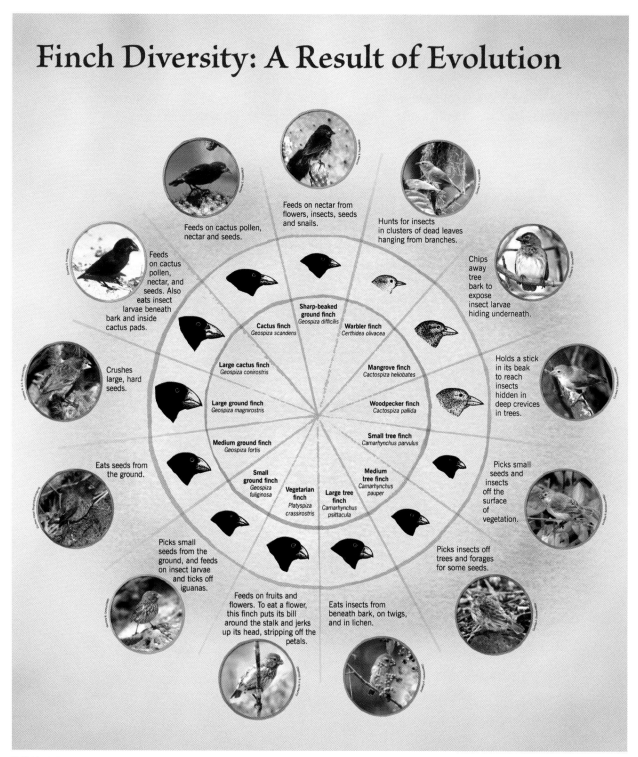

Feeds on cactus pollen, nectar and seeds.

Feeds on nectar from flowers, insects, seeds and snails.

Hunts for insects in clusters of dead leaves hanging from branches.

Feeds on cactus pollen, nectar, and seeds. Also eats insect larvae beneath bark and inside cactus pads.

Chips away tree bark to expose insect larvae hiding underneath.

Crushes large, hard seeds.

Holds a stick in its beak to reach insects hidden in deep crevices in trees.

Eats seeds from the ground.

Picks small seeds and insects off the surface of vegetation.

Picks small seeds from the ground, and feeds on insect larvae and ticks off iguanas.

Picks insects off trees and forages for some seeds.

Feeds on fruits and flowers. To eat a flower, this finch puts its bill around the stalk and jerks up its head, stripping off the petals.

Eats insects from beneath bark, on twigs, and in lichen.

Sharp-beaked ground finch *Geospiza difficilis*

Cactus finch *Geospiza scandens*

Warbler finch *Certhidea olivacea*

Large cactus finch *Geospiza conirostris*

Mangrove finch *Cactospiza heliobates*

Large ground finch *Geospiza magnirostris*

Woodpecker finch *Cactospiza pallida*

Medium ground finch *Geospiza fortis*

Small tree finch *Camarhynchus parvulus*

Small ground finch *Geospiza fuliginosa*

Vegetarian finch *Platyspiza crassirostris*

Large tree finch *Camarhynchus psittacula*

Medium tree finch *Camarhynchus pauper*

Color Plate
6

Our Closest Living Relatives

DNA reveals that humans share a common
ancestor with apes, our closest living relatives.

Chimpanzee
Pan troglodytes

Bonobo
Pan paniscus

Human
Homo sapiens

Gorilla
Gorilla gorilla

Orangutan
Pongo pygmaeus

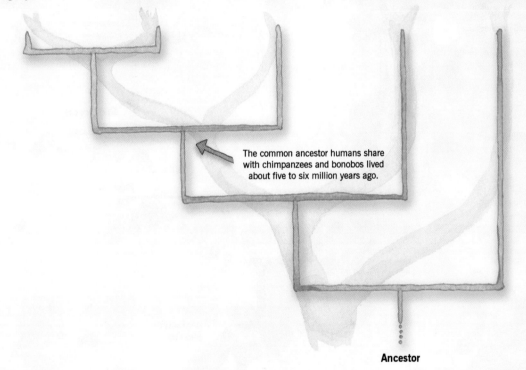

The common ancestor humans share
with chimpanzees and bonobos lived
about five to six million years ago.

Ancestor

Whales: Evolution from Land to Sea

Today's whales evolved from four-legged land mammals that lived about 55 million years ago.

Hippopotamus
Hippopotamus amphibius

Baleen whales
Balaenoptera musculus
(blue whale)

Toothed whales
Tursiops truncatus
(Atlantic bottlenose dolphin)

Today

35 million years ago

Elomeryx

36 million years ago

Dorudon

Rodhocetus

47.5 million years ago

Pakicetus

48.5 million years ago

55 million years ago

Artiodactyl ancestor

SMM illustration, ancient Whales John Klausmeyer illustration, modern Whales and Hippo Adam Wiens illustration.

Virus and the Whale: Exploring Evolution in Creatures Small and Large

Chapter 2

Evolution in Seven Organisms

Carl Zimmer

In the activities section of this book, you will be able to explore seven examples of evolution and learn how scientists are discovering some of the evolutionary history of each organism. Each case highlights an important aspect of evolutionary biology, from the "arms race" between viruses and their hosts to the long-term evolutionary changes that can turn a land mammal into a whale. Here I will examine some of the most important points of each case.

Viruses

It is debatable whether viruses are truly living organisms. They have genes and proteins like living organisms, they can replicate like living organisms, and they can even evolve like living organisms. But viruses typically can only replicate by taking over the cellular machinery of an animal, a plant, a bacteria, or some other organism. Despite this debate, however, viruses still represent some of the most striking examples of evolution in action.

Human immunodeficiency virus, or HIV, is the cause of acquired immunodeficiency syndrome (AIDS), one of the world's most devastating diseases. Some 45 million people are infected with it today, and without treatment most of its victims die. In some particularly hard-hit countries, HIV has cut decades off

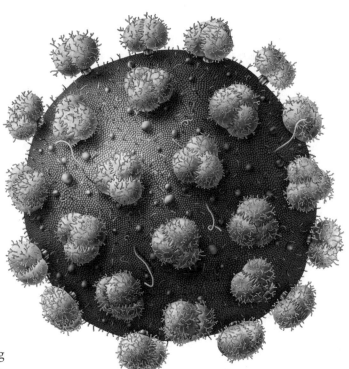

HIV, the Human Immunodeficiency Virus, causes the disease AIDS.
UNSM Angie Fox illustration

of the average life expectancy. To fight HIV, scientists must understand it. And to understand it, they must understand its evolution.

Like other viruses, HIV is composed of a small amount of genetic material encased in a protein shell. It invades a type of immune cell, called a T cell, and hijacks the machinery of the T cell to make more copies of itself. These copies escape the host cell to invade new T cells, and in some cases these new copies leave a victim's body altogether and infect another person. HIV is primarily spread through sexual contact, contaminated needles, and mother's milk.

Viruses face a ruthless foe in the form of the human immune system. When the immune system recognizes a pathogen, it mounts a devastating attack. Depending on the nature of the infection, immune cells may engulf the invaders, or they may produce toxic chemicals that latch onto the invaders.

The crucial step in fighting infections is recognizing them. Immune cells produce receptors and other molecules with a wide range of shapes. A few of these molecules may have a shape that allows them to latch onto an invading pathogen. Often they attach to a protein on the pathogen's surface. When an immune cell experiences this sort of success, it reproduces rapidly, producing an army of immune cells, all of which recognize the same pathogen. In the case of HIV, the human immune system tends to recognize a surface protein called Env. Once it makes this recognition, the immune system can seek out the viruses and destroy millions of them a day.

The immune system creates just the sort of conditions that favor extraordinarily rapid natural selection. As infected cells produce new copies of HIV, they also produce a high level of mutant viruses. In many cases, these mutants do a worse job of replicating than their ancestors and die off. But a few of them produce surface proteins with a slightly different structure. If the immune system is already primed to recognize its ancestor's surface proteins, the new virus can escape destruction. It will invade more cells than other HIV viruses and produce more copies of itself. With enough time, the immune system can recognize these new surface proteins, but the viruses can evolve still more proteins and continue to escape destruction.

Diatoms

Long before the modern science of biology emerged, people understood that animals and plants were not all alike. They came in many different kinds, which became known as species. Before the 1800s, species were believed to have been created in their current form, either at the beginning of the world or at some later time. But whenever a species was created, it was believed to be completely distinct from other species. It only reproduced with others of its kind and had a unique appearance distinguishable from other species.

Charles Darwin noted that the closer one looks at the natural world, the blurrier the boundaries between species become. In an individual species, for example, populations in different parts of its range may have different appearances. What's more, many species are able to breed with other species. In fact, in some cases it is hard to say where one species stops and another begins. Darwin argued that species were blurry because they were not created distinct from one another. Instead, old species gave rise to new species through gradual evolutionary change.

Darwin offered only sketchy ideas about how new species formed (a process called speciation). Today, scientists have amassed much more evidence about this process. Some of the evidence comes from the DNA of different populations of the same species, or of closely related species. Other lines of evidence come from breeding experiments. And still other lines of evidence come from the fossil record.

While many aspects of speciation remain to be fully understood, a coherent picture of the process has emerged. One of the most important ways for new species to form is isolation. A species is typically made up of a

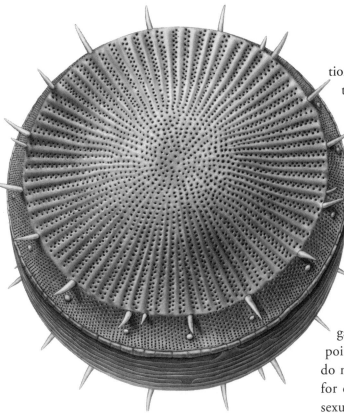

The diatom, *Stephanodiscus yellowstonensis.*
UNSM Angie Fox illustration.

tion becomes more and more distinct. If it then comes back in contact with the rest of their species, it may now be impossible for them to interbreed. It's also possible that they can produce hybrid offspring, but hybrids that are much less successful at reproducing than their parents. The members of the isolated population can now only breed successfully among themselves. They have, in other words, become a new species.

This model does a good job of explaining how sexually reproducing organisms form new species. It should be pointed out, however, that many organisms do not have to reproduce sexually. Bacteria, for example, simply divide in two. In nonsexual organisms, a species may simply be a lineage of organisms that has undergone so much natural selection that it is recognizably distinct from other species.

In Activity 2 (Diatoms: One-Celled Wonders) we will look at an example of the sort of evidence scientists examine to understand how new species evolve. The case involves diatoms, which are aquatic single-celled organisms that harvest the energy from sunlight like plants do. Recently researchers discovered a new species of diatom in Yellowstone Lake, which they dubbed *Stephanodiscus yellowstonensis.* Scientists can recognize different species of diatoms by studying the ribs and spines that ornament their plates. Each species has a distinct pattern of these ornaments, with a little variation from one individual to the next. *Stephanodiscus yellowstonensis* has a distinctive pattern that scientists have not seen in diatoms in other lakes in the region. It appears then that the species lives only in Yellowstone Lake.

number of populations. The populations in different parts of the species's range may face different challenges for survival. For example, the northernmost members of a species may have to withstand much colder winters than the southernmost ones. Natural selection acts on the variation within the populations, making some genes more common than others. If the populations remain connected to one another, individuals will interbreed, spreading these evolved genes through the species.

But sometimes a population gets isolated from the rest of its species. A few birds may get swept away to a distant island, or a mountain range may gradually rise up to divide a species in two. Now the isolated population may evolve a number of new genes, which do not spread to the rest of the species. The popula-

To understand the evolution of this unique species, researchers dug into the muddy lake bottom and found a continuous record of diatom fossils reaching back 14,000 years, to when the lake first formed. The oldest fossils have a pattern of ribs and plates that's outside the variation of today's *Stephanodiscus yellowstonensis.* Instead, they are more similar to another species, *Stephanodiscus niagarae*, which is found today in neighboring lakes. Over the course of 4,000 years, the fossils show that the diatoms of Yellowstone Lake gradually changed, until about 10,000 years ago they took on the distinctive look of *Stephanodiscus yellowstonensis.* It is possible that the diatoms were experiencing natural selection due to a warming climate. The lake may have also been isolated enough from other lakes that the diatoms evolved into a distinct species of their own.

Ants and Fungi

Among the common misperceptions about evolution is that it has produced a world of brutal competition—akin to what Alfred Tennyson famously called "Nature red in tooth and claw." But many species exist in partnerships, depending on other species for their very survival. Many flowers depend on bees and other insects to spread their pollen to other flowers; in return, the insects rely on the flowers for nectar and other food.

These partnerships can be wonderfully complex. In Madagascar, for example, there is a species of moth with a tongue that measures almost a foot long when it is fully extended. Its long tongue allows it to feed on an orchid in which the nectar sits at the base of a flower almost a foot deep. As the moth nuzzles against the top of the flower to reach the hidden nectar, the orchid's pollen becomes attached to its head. In Darwin's lifetime, only the orchid was known, but he predicted that the amazing moth would someday be found. He was right, although the moth would not be discovered until 21 years after his death.

Darwin recognized that natural selection could produce these intimate partnerships—a process now known as coevolution. In Activity 3 (Ants & Co.: Tiny Farms) we will look at how scientists can learn about coevolution by studying ants that grow a fungus crop.

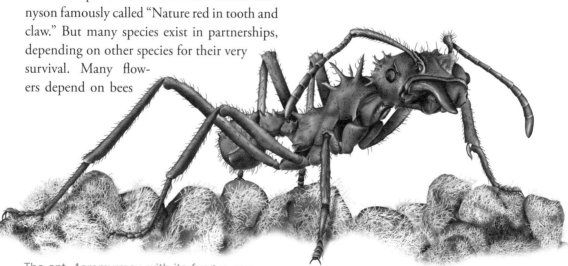

The ant, *Acromyrmex*, with its fungus crop.
UNSM Angie Fox illustration.

Over 200 species of ants in Central and South America are known to raise fungus crops in their nests. The ants fertilize the fungus crop with leaves and other plant matter, which they harvest from the surrounding forest and bring into their nests, where they chew it into a paste. The fungi grow on the plant paste, and the ants eat the fungi.

These partners have coevolved into an intimate interdependence. The fungi cannot survive outside of the ant nests, and they produce special nutrient-rich packages for the ants to eat. The ants in turn cannot survive without the fungi. When a queen ant leaves her birth nest to found a new one, she makes sure to take a mouthful of fungi with her to start a garden of her own.

Recently, researchers have discovered that two other partners are part of this remarkable coevolution. One is a parasitic form of fungus that specializes in attacking the garden fungi. The parasite can wipe out an entire crop, which raises the question of why so many nests are healthy. Researchers discovered that the ants have special cavities on their bodies that serve as homes for a special form of bacteria. The bacteria produce antibiotics that kill the parasitic fungi—but not the fungus crop.

Researchers are reconstructing the coevolutionary history of these partners by comparing the different species of ants, fungi, and bacteria. By analyzing the DNA of the ants, for example, they have constructed an evolutionary tree that illustrates how the species are related to one another. They have also constructed the evolutionary tree of fungus crops and have lined up the two trees. It turns out that the pattern of branches in the fungus tree strongly resembles the pattern in the ant tree. That suggests that as ants branch into new species, the fungi that they raise branch into new species as well.

Flies

In this chapter, we've looked mainly at the role of natural selection in evolution. But life can change in other ways. Reproductive success depends not only on withstanding the elements and finding enough food to survive. An individual also has to be able find a mate. If females are more attracted to one male than another, the more successful male will father more offspring—offspring that will carry his genes. Darwin first recognized this alternative to natural selection and termed it sexual selection.

Much of the diversity in animals has emerged thanks to sexual selection. In some species, males compete with one another to be able to mate with females, and they have evolved elaborate weapons for the fight. Mountain sheep have massive horns that they use to ram each other. Male northern elephant seals fight by slamming their huge bodies against one another. The bigger the seal, the more likely he is to win. As a result, male elephant seals can weigh up to 2,700 kilograms, three times more than females.

On the other hand, males have also evolved elaborate traits that serve only to attract females. Male peacocks display their resplendent tails so as to catch the eyes of potential mates. Female peahens are very tough critics of peacock tails. They prefer to mate with males with larger, more elaborate tails with more iridescent "eyes." Other female birds judge prospective mates by their songs; the more complex the song, the more likely a male is to find a mate.

In Activity 4 (Hawaiian Flies: Song & Dance Success), we will look at a spectacular case of sexual selection at work.

Hawaii is home to more than 800 species of flies in the genus *Drosophila* (the same genus that includes the common "fruit fly" in the United States). The Hawaiian *Drosophila* are unusual not only because there are so many species found on such a small island chain, but because these species are found nowhere else on Earth. By studying the natural history and the DNA of these flies, scientists have been able to reconstruct their evolutionary history. Drosophila flies settled on Hawaii several million years ago. Because the islands are so remote from continents, they rarely receive such visitors. As a result, the colonists did not face competition from well-established flies. They were able to adapt to many habitats and form new species. Another factor that led to additional species was the geology of Hawaii. The islands are a chain of undersea volcanoes formed as the crust slides over a giant mass of hot rock. Every few hundred thousand years a plume of this rock pumps lava onto the ocean floor, forming a new island that eventually rises above sea level. Flies on neighboring islands have colonized these new islands and then evolved into new species.

Drosophila flies—both in Hawaii and the rest of the world—carry out long courtship behaviors before mating. Males will produce distinctive buzzing songs and fly around females in dance-like patterns. Researchers have found that despite the rapid rate at which new species form in Hawaii, sexual selection has also kept up the same pace. Each new species has its own distinct mating rituals. Sexual selection may have even played a pivotal role in the formation of new species. Researchers have suggested that when a population of Hawaiian flies became separated from the other members of their species, they could only mate among themselves. As they adapted to their new environments, they also evolved new mating rituals—including new dances and songs the males performed for females. The females became attracted to new features on the bodies of the males, such as wing patterns or big antennae. Over thousands of generations, these flies became unable to breed with their ancestral species. In other words, they had branched off into a new species.

Two male Hawaiian flies, *Drosophila heteroneura*, in the head-to-head posture of their territorial defense display.
UNSM Angie Fox illustration.

The medium ground finch,
Geospiza fortis.
UNSM Angie Fox illustration.

Finches

Charles Darwin believed that natural selection acted so slowly, over the course of so many generations, that its effects could only be studied indirectly. Many generations of biologists followed Darwin in this belief. While it might be possible to observe natural selection in the laboratory, watching natural selection in action in the wild was more than could be hoped for.

That has changed in large part thanks to the work of Peter and Rosemary Grant, a husband-and-wife team of biologists at Princeton University. For 30 years they have studied finches that live on the Galápagos Islands in the eastern Pacific. These birds were first studied scientifically by Charles Darwin, and they've since been dubbed Darwin's finches. Darwin's finches provide an excellent opportunity to document natural selection. On a given island, a few hundred birds may hatch in a given year, and most of the birds spend their entire lives there. It is thus possible to measure every single bird on an island. That allows the Grants to document many kinds of variation in a finch population, from their body mass to the width of their beaks. The Grants can also compare individual finches to their offspring, to determine how strongly inherited these variations are. The islands also undergo intense swings in weather, with periods of droughts alternating with heavy rains. These swings alter the food supply, which can create conditions for strong natural selection to take place.

The Grants have made some of the most striking measurements of natural selection on medium ground finches that live on the island of Daphne Major. The medium ground finches use their heavy beaks to crack seeds of plants that grow on the island. *Chamaesyce amplexicaulis* produces small seeds, while *Tribulus cistoides*, commonly called caltrops, produce seeds encased in hard, woody fruits. Finches with big beaks (11 millimeters deep) can crack open the fruit in 10 seconds. Finches with beaks 10.5 millimeters across need 15 seconds. And finches with beaks 8 millimeters deep give up on caltrops. Instead, they stick to eating small seeds.

A 1977 drought killed off most *Chamaesyce amplexicaulis* plants, leaving the medium ground finches without any small seeds to eat. Many birds died because they couldn't crack open caltrop fruit. The Grants discovered that within a few years the population of finches recovered. But now their beaks were on average 4% deeper (about half of a millimeter). That's because finches with big beaks had a better chance of surviving the drought.

At the end of 1982, heavy rains came to the islands. *Chamaesyce amplexicaulis* bloomed, producing lots of small seeds. The small-beaked birds could eat the small seeds more efficiently than the big-beaked ones, and so now natural selection favored them. As a result, the average size of beaks decreased by 2.5% (about a fifth of a millimeter). In other words, the Grants have documented natural selection at work over the course of just a few years.

Since the Grants pioneered the measurement of natural selection in wild populations, other researchers have also documented it in other species. Natural selection is happening all around us; we can see it if we know how to look for it.

Humans

When Charles Darwin was formulating his theory of evolution by natural selection, he spent a great deal of time thinking about the origins of human beings. He recognized that humans display all of the elements necessary for evolution. Individual humans vary in their traits. Some of these traits get carried down from one generation to the next, and they can influence how many children individuals can have. Natural selection should favor the traits in humans that boost the reproductive success in a given environment.

You can also find evidence that humans are the product of evolution by looking at the human body itself. Our bodies are very similar to those of monkeys and apes. Indeed, almost a century before Darwin published *On the Origin of Species*, the naturalist Carl Linnaeus placed humans in the Primate family. And our bodies contain many traits that can only be explained as vestiges of our evolutionary history, such as the coccyx, the section of the spine that extends below the pelvis. Anatomical compari-

sons show that it is actually a vestigial tail.

Darwin hypothesized that humans and other primates shared a common ancestor. He speculated about how even human behavior might be the product of evolution. He paid regular visits to the London Zoo to observe an orangutan named Jenny, taking notes about the similarity of her facial expressions to those of humans.

For years Darwin kept his ideas about human origins to himself. He filled *On the Origin of Species* with enough animals to stock the London Zoo many times over. Whales, bears, and flatworms parade through his pages, along with giant extinct sloths, armadillos, and marine reptiles. But one species is glaringly absent—*Homo sapiens*. "Light will be thrown on the origin of man and his history," was all he would say.

Darwin knew that it would be hard enough to persuade his readers that animals or plants evolved. To add humans to evolution's list of accomplishments might instantly turn them away. "I thought that I should thus only add to the prejudices against my views," he later wrote.

He was probably right. Most of the attacks launched against *On the Origin of Species* in the 1860s sooner or later came around to the question of man's place in nature. Sir Richard Owen, the greatest British anatomist of the 19th century, thought he could refute evolution by finding a part of the brain that was unique to humans. Such a trait would set humans safely apart from other apes, and thus out of the grasp of Darwin's natural selection. In 1860, Owen attended the most famous debate over Darwin, at Oxford University. Bishop Samuel Wilberforce delivered a furious attack on *On the Origin of Species*, and then turned to Thomas Huxley and asked him whether he descended from an ape on his mother's or father's side.

As the years passed, however, the attacks faded. Naturalists still debated about how evolution worked. Did life have a built-in direction along which it evolved, they wondered, or was it nothing more than the adaptation of populations to the randomly changing conditions they faced? But all sides increasingly agreed that life had indeed evolved. Seeing a consensus emerging, Darwin decided the time was right to put humans in his zoo. In 1871 he published *The Descent of Man*, presenting evidence that humans had evolved and suggesting that they were modified mammals descended from an ape-like ancestor.

Today, scientists have far more evidence at hand than Darwin had to test ideas about human evolution. In many ways he has been splendidly vindicated. Fossils of hominids have been found dating as far back as six million years. For the first four million years of hominid ancestry, all known species lived in Africa. But scientists have also discovered many things that would surprise Darwin. For example, Darwin thought that as the ancestors of humans evolved bipedalism, they evolved large brains at the same time. But fossils reveal that the first hominids to stand upright had brains one-third the size of a modern human's. It took three million years before their descendants would evolve a brain approaching our own.

Traditionally, scientists have looked to fossils and comparisons with other primates for clues to human evolution. But human DNA turns out to hold a great deal of information about our evolution, information that is only beginning to emerge. In Activity 6 (Humans

Only a few percent of the DNA of humans and chimps is different.
UNSM Angie Fox illustration.

& Chimps: All in the Family), we will consider the research of German researchers Svante Pääbo and Henrik Kaessmann, who study the relationship of humans to living apes. Humans have DNA more like living great apes—orangutans, gorillas, bonobos, and chimpanzees—than like any other living animals. And among the apes, our DNA shows the clearest kinship with chimpanzees. In fact, many stretches of DNA from one species are identical, nucleotide for nucleotide, with the other.

Pääbo, Kaessmann, and other researchers are currently sorting through the human genome to determine which portions are identical to chimpanzees and which are unique to our own lineage. It is this human-specific genetic material where scientists may discover the molecular history of our species.

Whales

While DNA may provide a new set of clues about evolution, fossils remain a vital source of information about the history of life—information that would otherwise be impossible to discover. The evolution of whales is a case in point.

Before the 1980s, whales represented one of evolutionary biology's most fascinating open questions. Superficially, whales (including dolphins and porpoises) seem like fish. They spend their entire lives in the water. They have flippers and lack hind legs. Their bodies are streamlined, and they generate lift with their tails in much the same way sharks or fish do.

And yet whales share many traits with land mammals—traits that fish lack. As we mentioned earlier, whales produce milk. They have placentas and give birth to live young. What's more, underneath their fish-like bodies, they have bones, muscles, and various or-

gans that can be found in mammals but not in fish. While fish breathe primarily through gills, whales have lungs, which take air in through the blowhole on top of their skull.

Darwin and other 19th century evolutionary biologists recognized that this puzzling mix of anatomy pointed to an extraordinary history for whales: Their ancestors were land mammals that gradually adapted to life in the water. Their mammalian bodies were modified in the process, giving them a superficially fish-like anatomy.

This hypothesis raised a number of questions. Which group of mammals gave rise to whales? Where did the transition take place, and when? How long did the transition take, and what did the intermediate forms look like? It would be long after Darwin's death before paleontologists would begin to find some clues.

Philip Gingerich of the University of Michigan has found a number of these transitional whale fossils. One of the most important of these fossils, *Rodhocetus*, lived 47.5 million years ago in Pakistan. Its anatomy was in many ways intermediate between land mammals and living whales. It had short limbs that it could use to swim in the water, but could have used on land only to drag its body around. Its nose was midway up its skull, no longer a typical mammal nose but not yet a blowhole. Although *Rodhocetus* looked like no whale alive today, it has a set of traits found only in modern whales, such as a distinctive bony case around its ears.

Gingerich and other researchers have discovered other fossils of whales that were more terrestrial, and later fossils that were more adapted to water. By 40 million years ago, 50-foot-long whales were swimming through the oceans, having completed the shift to their new habitat. But they still retained tiny, fully-formed hind legs as a vestige of their life on land. Enough fossil whales have been discovered now to teach us some important lessons about how evolution works. One of the most important lessons is that a dramatic transformation does not have to occur in a single leap. Instead, a lineage of animals can pass through a series of intermediates. Living whales may be a puzzle, but fossils of their ancestors show that they are just one more ever-evolving branch of the tree of life.

Teaching and Learning About Evolution

E. Margaret Evans

The National Association of Biology Teachers considers evolution to be the foundation for middle school life science. In the National Science Education Standards (NSES), evolution is an essential component of the science curriculum at all grade levels. With this book as your guide, your charge is to help youth learn about evolution as the unifying theme of the life sciences. How do you guide kids' understanding? What are the potential pitfalls? This chapter covers what you need to know based on the most recent research and advice about the teaching and learning of evolution. First, I shall present the National Science Education Standards. Then I will turn to current research on some of the best ways to teach evolution. Finally, I will tell you what researchers know about how kids learn about evolution and ways to use this knowledge when you teach, whether in informal or formal settings.

All major scientific and research organizations in the United States agree that evolution is a major unifying concept in the life sciences and should be included in the K–12 science education frameworks and curricula (NSTA 2003). Why is there such consensus? Scientists use the theory of evolution because it explains, with simplicity and elegance, the similarities and diversity found in all living organisms.

Whether today's youth plan to be scientists, health practitioners, teachers, engineers, politicians, or informed citizens, they must grasp the unifying role of evolutionary theory in 21st-century life science.

Teaching biological evolution can be a challenge. Our everyday intuitions about the way the world works are at odds with an evolutionary perspective. Instead of a static world inhabited by separate living kinds, evolutionary theory provides us with a dynamic world in which all living kinds are related, through a common ancestry. Evolutionary science provides a new way of thinking about the living world.

National Science Education Standards for Biological Evolution

Leading educators and scientists have come together to develop a set of national science standards aimed at improving the scientific literacy of the U.S. population. There are two major organizations involved in this effort, and although they have each tackled the standards from a different perspective, they have achieved a remarkable consensus on what students at different grade levels need to know about science. The National Research Council (NRC) developed the National Science Education

Standards, whereas the American Association for the Advancement of Science (AAAS) created the Benchmarks for Science Literacy and its more dynamic form, the *Atlas of Science Literacy.* According to these documents, an understanding of biological evolution and the nature of science should be a core component of science curricula for all students.

I shall briefly summarize the current consensus on standards for evolution content and the nature of science content for each grade level, based mostly on the NSES. The AAAS Benchmarks break down the grade levels somewhat differently (K–2, 3–5, 6–8, and 9–12), but in general the concepts expected of each age group are similar (see Table 1). The distinctive feature of the Atlas of Science Literacy is that the learning goals for a particular topic are presented on one page as a conceptual flowchart from the earlier to the later grade levels (see the AAAS flowchart for biological evolution on page 27).

Table 1: *Atlas of Science Literacy*: Learning Goals

- Scientific Inquiry: Evidence and Reasoning in Inquiry
- Scientific Inquiry: Scientific Investigations
- Scientific Inquiry: Scientific Theories
- Scientific Inquiry: Avoiding Bias in Science
- Heredity: DNA and Inherited Characteristics
- Heredity: Variation in Inherited Characteristics
- Cells: Cell Functions
- Evolution of Life: Biological Evolution
- Evolution of Life: Natural Selection

(AAAS 2001)

Evolution and the Nature of Science in the Classroom:
A grade-level comparison

Below is a summary of the relevant sections in the NSES, for each range of grade levels. A chart detailing the relationship between each of the activities in the book and the National "Life Science" Standards (Content Standard C) for Grades 5 to 8 can be found in this book on pages xvii–xix. Further information is provided in the resources section at the end of the book.

Kindergarten to fourth grade

The NSES propose that children in this age-range be taught about the characteristics of different kinds of organisms (both plants and animals), their life cycles, and their ecological niches (see Table 2). In particular, they learn about the interrelationships between living things through direct experience of the living world. Young children are introduced to the nature of science as they carry out simple experiments and learn basic concepts about science as a way of thinking about the natural world. These activities are intended to be the foundational knowledge base for an understanding of the life sciences and of evolution.

Table 2: National Science Education Standards in the K–4 Classroom

Content Standard A: Science as Inquiry

Content Standard C: Life Science

Content Standard F: Changes in Environments

Content Standard G: History and Nature of Science

(NRC 1996)

Evolution of Life: Biological Evolution (AAAS 2001)

12

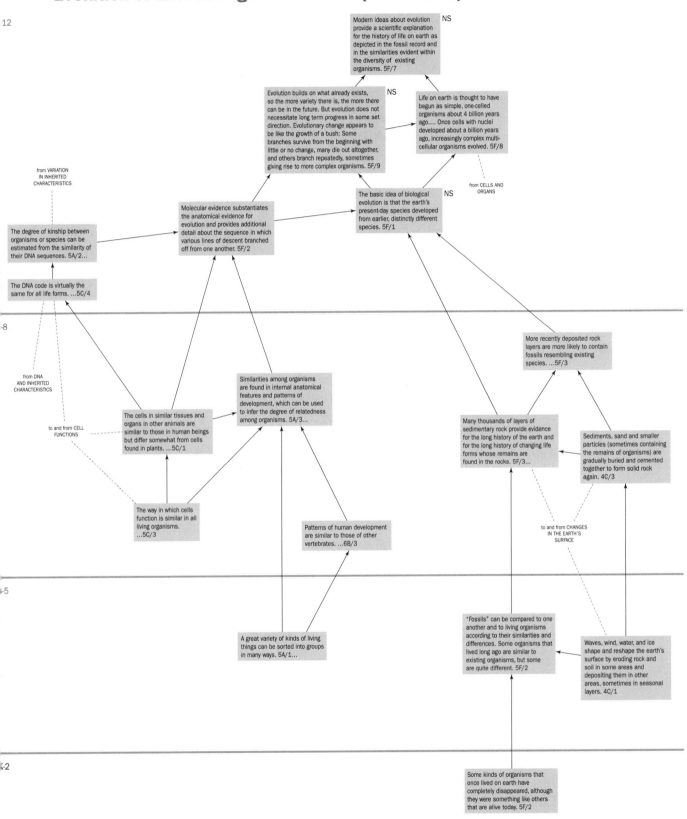

Modern ideas about evolution provide a scientific explanation for the history of life on earth as depicted in the fossil record and in the similarities evident within the diversity of existing organisms. 5F/7 — NS

Evolution builds on what already exists, so the more variety there is, the more there can be in the future. But evolution does not necessitate long term progress in some set direction. Evolutionary change appears to be like the growth of a bush: Some branches survive from the beginning with little or no change, many die out altogether, and others branch repeatedly, sometimes giving rise to more complex organisms. 5F/9 — NS

Life on earth is thought to have begun as simple, one-celled organisms about 4 billion years ago.... Once cells with nuclei developed about a billion years ago, increasingly complex multi-cellular organisms evolved. 5F/8

from VARIATION IN INHERITED CHARACTERISTICS

from CELLS AND ORGANS

The degree of kinship between organisms or species can be estimated from the similarity of their DNA sequences. 5A/2...

Molecular evidence substantiates the anatomical evidence for evolution and provides additional detail about the sequence in which various lines of descent branched off from one another. 5F/2

The basic idea of biological evolution is that the earth's present-day species developed from earlier, distinctly different species. 5F/1 — NS

The DNA code is virtually the same for all life forms. ...5C/4

8

More recently deposited rock layers are more likely to contain fossils resembling existing species. ...5F/3

from DNA AND INHERITED CHARACTERISTICS

Similarities among organisms are found in internal anatomical features and patterns of development, which can be used to infer the degree of relatedness among organisms. 5A/3...

to and from CELL FUNCTIONS

The cells in similar tissues and organs in other animals are similar to those in human beings but differ somewhat from cells found in plants. ...5C/1

Many thousands of layers of sedimentary rock provide evidence for the long history of the earth and for the long history of changing life forms whose remains are found in the rocks. 5F/3...

Sediments, sand and smaller particles (sometimes containing the remains of organisms) are gradually buried and cemented together to form solid rock again. 4C/3

The way in which cells function is similar in all living organisms. ...5C/3

Patterns of human development are similar to those of other vertebrates. ...6B/3

to and from CHANGES IN THE EARTH'S SURFACE

5

A great variety of kinds of living things can be sorted into groups in many ways. 5A/1...

"Fossils" can be compared to one another and to living organisms according to their similarities and differences. Some organisms that lived long ago are similar to existing organisms, but some are quite different. 5F/2

Waves, wind, water, and ice shape and reshape the earth's surface by eroding rock and soil in some areas and depositing them in other areas, sometimes in seasonal layers. 4C/1

2

Some kinds of organisms that once lived on earth have completely disappeared, although they were something like others that are alive today. 5F/2

evidence from existing organisms

fossil evidence

Fifth to eighth grade

For these grade levels, the NSES explicitly mention biological evolution. Children are encouraged to pay attention to the relationship between the organism and its environment, its ecosystem, and, in particular, the concepts of diversity and adaptation. They are taught about the function and structure of cells. Children also learn that how any one species moves, reproduces, and gets food is a function of its evolutionary history. The concepts of variation, inheritance, selection, and time are introduced, along with the fossil record and extinction. These concepts are also woven into a larger context—that of the study of systems—with the history of Earth and the universe portrayed as tightly coupled systems. Students' classroom activities in the fifth through eighth grades include studies of the nature and history of science. These students learn about observation, experimentation (hypothesis testing), the relationship between explanation and evidence, and modeling, particularly of theoretical and mathematical models (see Table 3).

Table 3: National Science Education Standards in the 5–8 Classroom

Content Standard A: Science as Inquiry

Content Standard C: Life Science

Content Standard D: Earth in the Solar System

Content Standard G: History and Nature of Science

(NRC 1996)

Ninth to twelfth grade

NSES for these grade levels detail the Darwinian concepts of natural selection and common descent, along with gene theory and the molecular basis of heredity. Students learn about biological classification as a hierarchy, determined by the evolutionary relationships between organisms, with species as the fundamental biological unit. Biological evolution is again woven into a larger context along with that of the origin and evolution of Earth and the universe. These students study science as a special way of knowing, based on empirical standards, logical arguments, explanation, and skepticism. They discover the principle that all scientific knowledge is subject to change in the light of new evidence (see Table 4).

Table 4: National Science Education Standards in the 9–12 Classroom

Content Standard A: Science as Inquiry

Content Standard C: Life Science

Content Standard D: The Origin and Evolution of the Earth System

Content Standard G: History and Nature of Science

(NRC 1996)

Summary

The NSES present the theory of evolution as a unifying theme for life science. Evolution integrates a life science curriculum focused on biological systems that are interrelated and constantly undergoing change. Change in one part of the system affects the other components. The origin of biological organisms is also presented as part of a broader topic, the origin of the Earth and of the universe. Importantly, science is presented as an activity that is based on evidence derived from meticulous investigations of the natural world. Over the grade levels these topics are gradually introduced, with the requirements for each level offering more depth, while building on early foundational concepts.

Teaching Science

The National Science Education Standards for teaching advocate inquiry-based activities as the best method for learning about evolution (NAS 1998; NRC 2000). The goal of the inquiry method is to have the student think and behave like a scientist. This goes beyond hands-on activities, in which a student may engage in some aspect of the scientist's behavior, such as collecting data or exploring natural phenomena. Inquiry-based learning emphasizes thinking and reasoning as well. It is known as the "minds-on" approach: Engage, Explore, Explain, Elaborate, and Evaluate (the 5E model). Students have to come up with hypotheses to explain a pattern of observations, and then conduct an experiment or a study designed to test these hypotheses. Once they have gathered the data, they have to explain the results, decide whether or not they fit with the original hypotheses, and consider alternative explanations. Finally, they have to explain their results to an audience, either in a written or an oral form. There are many data sets available that can be used to explore evolutionary hypotheses. Moreover, all of the activities in this book incorporate basic concepts of inquiry-based learning.

In contrast to the more traditional approach, the inquiry method also emphasizes collaborative learning. It is the rare scientist who works entirely alone. He or she would not be able to accomplish the work without a team. Lectures, memorization, and individual problem sets, which are the hallmarks of the traditional approach, are not absent with the inquiry method, but they clearly play a less dominant role. Overall, the inquiry method emphasizes the dynamic aspect of a scientific investigation. Science is viewed as a dynamic enterprise with current facts and theories acting as a way station to new facts and theories. Science learning experiences, in any setting, focus less on the accumulation of many superficial facts and more on the deep learning of fundamental principles.

How does inquiry-based or experiential learning play out in the everyday activities of the scientist's laboratory? Peter Medawar, a well-known evolutionary biologist, described some of the conversations in his laboratory:

- "What gave you the idea of trying...?"
- "What happens if you assume that...?"
- "Actually your results can be accounted for on a quite different hypothesis...."
- "Obviously a great deal more work has got to be done before...."

According to Medawar's own observations and those of others who have studied scientists at work, what is happening is that the scientists are "building explanatory structures, telling stories which are scrupulously tested to see if they are stories about real life" (Hoagland 1990, p. 20). These "stories" are more formally represented as hypotheses, which are then tested by collecting data to determine their truth or falsity.

The 4-H Youth Development programs share the same basic approach as that of inquiry-based learning, but with a greater emphasis on experiential learning, the hallmark of 4-H activities. Their credo is "Do-Reflect-Apply," which involves youth carrying out an activity, reflecting on the experience, then applying this experience to their own lives or that of the broader society.

The Development of Children's Concepts of Evolution

Cognitive scientists are interested in the way people reason about the world, in the absence of expert training. Recently, there has been a surge of interest in this everyday reasoning and what impact it has on the developing learner's capacity to absorb new information. Commonsense or everyday reasoning has been portrayed as a limited series of intuitive theories about the world, each one of which potentially describes a different kind of knowledge. Intuitive theories differ from scientific theories in many ways, but like scientific theories they frame the way we view the world and provide both questions and explanations (Wellman and Gelman 1998). These are the kinds of everyday explanations or hunches that most easily come to mind when we try to figure out what is going on in the world. How these intuitive theories develop over the course of a child's life and into adulthood is the focus of much recent research. I'll focus on the development of those concepts most likely to impede evolutionary thinking. Family background and the development of children's intuitive theories are the major influences on children's reasoning about evolutionary change.

How do we know about these developmental changes? Over the past 15 years or so, my colleagues and I have begun to map out the emergence of children's understanding of natural transformations, such as evolutionary change, which relate to the development of their intuitive theories of biology (Evans 2001;

Rosengren, Gelman, Kalish, and McCormick 1991). In the process we have interviewed hundreds of children and their parents from different religious backgrounds. To investigate children's intuitive ideas, we ask the children unexpected questions and give them unusual tasks. This is done to make sure that they do not just give us rehearsed responses. They have to think through the issues. We also compare children's intuitions about natural transformations, such as seasonal change, with their intuitions about artificial transformations, such as making chairs or toys. These kinds of studies reveal the way children of different ages reason about different types of transformations. I shall detail what we have found and tie it in with the extensive research that science educators have carried out on older students' ideas about evolutionary change.

Four- to seven-year-olds

Over the preschool and elementary school years, children slowly abandon their idea that animals cannot change. In preschool through about second grade, most children reject the idea of almost any kind of radical biological change, from metamorphosis to adaptive variation. Thus, from the perspective of the young child's intuitive biology, living things cannot change. This age group is learning so much that is different and new, it is surely useful to have an intuitive sense that the world around them is permanent and enduring. Young children, however, do know that animals possess adaptive features such as wings for flying or fins for swimming, but they have little sense of what would happen if the environment changes. If you ask a child from this age group where the very first animals come from, you will get a variety of answers. Some are likely to respond that God made them. Other children may well reply that the very first animal came "from someplace else" or that they "came out of the ground." In other words, they appear to think that the animals were always here on Earth, but someplace else where they could not be seen. This idea may be rooted in children's everyday experiences of the world. Every spring after the snow melts or after the first spring rains, the ground seems to burst with new life.

Eight- to ten-year-olds

From about third grade to the end of fourth grade, there is a gradual shift in children's reasoning. This age group is more likely to accept some kinds of radical biological change, especially over the life cycle, such as metamorphosis. Interestingly, whatever their family background, most children in this age range endorse the idea that the very first kinds of animals were "made by someone," and often that someone is God. One reason for this belief is that unlike their younger siblings, older children are beginning to think about existential questions. This age group is more likely to know about death and understand that animal kinds are not eternal, in that they were not always here on Earth, nor will they continue to be on Earth. So, the question arises, how did different kinds of animals get here in the first place? These children appear to transfer their intuitive understanding of the human as an intentional manufacturer of new tools, and apply it to objects that have arisen naturally, such as new species.

Simultaneously, children in this age range are starting to integrate different kinds of causes into a complex causal structure. If preschool children see "Josh" knock over a glass and break it, they are perfectly capable of reasoning about the immediate cause. They can tell you who knocked over the glass and how it happened, such as "Josh didn't see the glass." But if you ask preschoolers to think a little harder about "why" Josh knocked over the glass, they have more difficulty. The older children, however, are better able to engage in a more complicated reasoning process and arrive at a more distant cause, as in the following causal chain: Josh knocked over the glass because he was in a bad mood, because he didn't get lunch, because he forgot his lunch money, and so on, until they arrive at the most distant or original cause. This sort of reasoning is necessary for understanding the origins of novel animal kinds: Why and how did something come into existence in the first place?

Ten- to twelve-year-olds

On the surface, at least, the beliefs of preadolescents are very similar to the beliefs of the adult members of their community, with the same percentage endorsing evolutionist and creationist beliefs. Children in fifth grade and older are able to reason about existential questions such as the origins and death of living things. At this point, we often see the influence of a family's system of beliefs. Children who are exposed to the evidence that animals change—from metamorphosis, to adaptive variation within species, to fossils—are most likely to accept major evolutionary changes. They will agree that one kind of animal could have originated from earlier and very different kinds of living things, although they are likely to exhibit many misconceptions about evolution. For these children who take the perspective of a naturalist, this is the beginning of a more complex understanding of the fundamental interrelationship between all living things. Conversely, children who know the least about natural history and fossils, and who go to schools that endorse Biblical literalism, are likely to endorse the idea that God created each kind of animal. But, interestingly, these beliefs seem to vary depending on the organism. Many children and their parents exhibit mixed beliefs, agreeing that butterflies and frogs evolve but that God created mammals, in particular humans.

Older youth and adults

Adolescents are often ready to assimilate basic evolutionary concepts. But their everyday intuitions continue to undermine the teaching of Darwinian theory. Most adolescents and even many adults endorse a pre-Darwinian theory of evolutionary change, which makes it difficult for them to grasp contemporary Darwinian concepts (e.g., Bishop and Anderson 1990). Perhaps the most revealing indicator of this kind of reasoning is the use of need-based or intentional explanations of evolutionary change: Animals change because they need to adapt to novel environments in order to survive. Such ideas appear to have their roots in children's and adults' understanding of the way humans fit in with or adapt to their environment. Additionally, many believe that such adaptations, acquired over the lifetime of the individual animal, can be inherited by future generations.

Education researchers have also found that science learners think of evolution as growth and improvement over time. Such ideas contribute to the rejection of the idea that living species are very likely to become extinct, which should be a core concept in major evolutionary change. Although many adults and children accept the extinction of the dinosaurs, they are less willing to generalize this understanding to include contemporary species, especially the human species (Poling and Evans 2004). Many have the idea that species continually adapt to new environments and do not really become extinct. Perhaps a more critical aspect of this problem is that the concept of extinction seems to arouse existential concerns. In practical terms, what this means is that older children and adults have difficulty contemplating the idea that humans and other species alive today might cease to exist.

Using Research on Learning to Teach Evolution

In this section, I'll show you how to identify different reasoning patterns among science learners. This will help you guide novice learners to a more informed viewpoint, and also help you assess their learning. The research described in this section is based on museum visitors' explanations of the evolutionary problems presented in this book (Evans et al. 2005).

The following reasoning patterns could serve as the basis of both your teaching and assessment tools.

- Informed Naturalistic Reasoners: People who propose Darwinian evolutionary explanations for the origin of species.
- Novice Naturalistic Reasoners: People who propose natural explanations, but who have little understanding of Darwinian evolutionary mechanisms. Many will use the pre-Darwinian concepts described earlier.
- Creationist Reasoners: People who propose supernatural explanations. In particular, they reference God's direct role in the origin of species.

- Mixed Reasoners: People who use more than one of the above reasoning patterns. This is the reasoning pattern found in most people.

By using these reasoning patterns as a diagnostic tool, you can give your science learners insights into the nature of their own reasoning. You cannot expect them to completely replace their intuitive ideas with Darwinian evolutionary concepts. But they might be able to construct dual frameworks. You will have succeeded when you and they recognize when they are shifting from informed to novice naturalistic reasoning patterns and back again. By reflecting on their own reasoning processes, they should be able to change their way of thinking about the problem.

Teachers and youth leaders can also use these reasoning patterns to assess learning. Ideally, any assessment of students' understanding should not be an endpoint in and of itself, but a further tool to help them consolidate what they have learned. Each of the activity chapters in this book has an assessment question (part four) in which the learner is asked to take on the role of a science reporter and explain a central problem in the evolution of each organism. These assessment questions provide learners with an opportunity to reflect on what they learned as they read the chapter introductions and carry out the various activities. This is a core component of inquiry-based and experiential methods. This process encourages the use of higher-order thinking skills. Students are not just memorizing facts, they are developing scientific explanations and reporting them to outsiders. They can do this individually, with the instructor evaluating each reporter's response, or as a group activity.

Informed Naturalistic Reasoners

The informed naturalistic reasoner uses evolutionary concepts and explanations rather than everyday intuitive reasoning to explain biological change. The core evolutionary concepts of variation, inheritance, selection, and time (VIST) could be used to assess children's responses to the questions at the end of each of the seven activities in this book. The VIST acronym from the University of California Museum of Paleontology website *(http://evolution.berkeley. edu)* provides a useful way of framing and remembering these concepts.

Variation

Variation refers to the differences among individuals in a population. These can be described as differences in a particular trait (feature or behavior), as a mutation, or as genetic differences. Here is an example of the kind of informed natural reasoning you might encounter about variation in Activity 5 about finch evolution (Evans et al. 2005): *The finches with the larger beaks survived I suppose. The ones who didn't died out....*

Inheritance

Inheritance refers to traits (factors) that are inherited and passed from one generation to the next. Students may not use the term inheritance, but just convey the idea that some factors are passed from one generation to the next. Here is an example of informed natural reasoning about inheritance from the same activity: *Those big-beaked finches were favored by the environment. So they were able to eat, breed, and then their offspring continued to do the same....*

Selection

Selection refers to the idea that organisms with traits that are adapted to the environment are more likely to survive (and pass these factors on to the next generation). A reasonable response might note the key environmental feature to which the organism is adapted. Here is an example of informed natural reasoning about selection in finches:

Well, the large-beaked birds were the only ones that survived because they could eat the seeds, and therefore they were the only ones that reproduced; and the ones with the small beaks lost out....

Time

The number of generations produced over a given time period is a clue to whether evolutionary change will occur rapidly (as in HIV) or slowly (as in whales). From one generation to the next a species may change ever so slightly, but given enough time, the result can be huge. Almost any reference to time acknowledges its crucial role.

Here is an example from the finch activity of a response that includes the four core evolutionary concepts—variation, inheritance, selection, and time (VIST):

Well, in that case I would assume that the birds evolved—well, the birds with the larger beaks were the ones better able to survive, since the larger beaks were more useful in getting the seeds. So that trait is the one that was selected for, and the birds that had the smaller beaks died out over time.... They didn't produce as many offspring.

Novice Naturalistic Reasoners

Novice naturalistic reasoners usually employ everyday reasoning, particularly intuitive ideas about biological change. We'll look at three typical kinds of everyday explanations used by science learners.

The first kind of explanation is called goal-directed or need-based change. Here is an example from the finch activity of a novice naturalistic reasoner's goal-directed response:

It's evolution. They had to—for survival, the beaks had to grow so the finch could eat. So they just adapted... their bodies adapted so that they could survive. That's not evolution, is it, it's another word. Is it development? Then their babies had those beaks.

This novice science learner is using everyday reasoning to explain how the finches could survive. She has recognized a need, which is that the finches do not have the right kinds of beaks to eat in this environment. The mechanism of change that she has identified is that the finch's beak must grow to meet the challenge. The term *development* captures an understanding that animals grow to meet an unmet future need, much in the way an individual animal develops from infancy to adulthood. The other intuitive idea expressed by this science learner is that this trait (the large beak) is acquired over the lifetime of the finch and can be passed on to future generations.

This kind of response is relatively sophisticated, but it isn't scientifically correct. What this novice science learner does not realize is that there is already a natural variation in beak size in this population and that those finches with the larger beaks are the ones that will survive. It would not take much to have her focus on the variation in the population and then figure out a Darwinian mechanism of change, such as natural selection.

The next explanation typical of novice naturalistic reasoning includes terms such as *thoughts, beliefs, wants, skills,* or *effort.* These terms are a sign that this science reasoner is using everyday intuitions about the way humans solve problems, called intentional reasoning. Here is an example from the finch activity of a novice naturalistic reasoner's intentional response:

Well, I tend to believe that a lot of animals…have capabilities of making adaptations. Like they wanted to increase the size of the beak to get the seeds so they tried to change their beaks to use in their daily life.

This kind of explanation is subtly different from the goal-directed reasoning described above. In using an intentional explanation, this science learner appears to assume that the finch can actively think about the problem and try to solve it by changing its beak, in the same way an athlete will actively train to achieve a better performance. Ask the science learner if the finch can really solve problems in the same way humans solve problems. They will probably recognize that this is unlikely. Then they can begin to think of alternative explanations for the change in beak size.

The third kind of novice naturalistic reasoning is called proximate (or immediate) cause reasoning. Recall that when asked about the origins of species, younger children would often respond as if the species were "always here on Earth, but someplace else." They did not address the most distant or original cause: Why and how did these new kinds of animals come into existence? Instead, the children described how they became visible. Interestingly, older youth and adults sometimes come up with similar immediate or proximate cause explanations to describe evolutionary change. They are most likely to do so when they are confronted with questions about the origins of organisms about which they know very little or which

are not visible, such as insects, diatoms, and viruses. Here are some examples from the diatom and HIV activities of a novice naturalistic reasoner's proximate cause response:

Diatom: *Water could have been mixed throughout the whole Earth, and it could have carried new algae in different places.*

HIV: *They were there and they weren't detected.*

There are several ways to address this kind of response. First, you should help the youth realize that there is a mystery. These particular organisms did not exist before and now they do. They were not hiding elsewhere on Earth. How did they come into existence? What are the relevant clues? What has changed? With some prodding the youth should begin to realize that the change in the environment is an important clue. Then they can make the connection between environmental change and the appearance of new species.

Creationist Reasoners

Religion and evolution are perfectly compatible, with a few exceptions. One exception is Biblical Literalism, whose adherents believe that God created each kind of animal that is currently on Earth just a few thousand years ago. Such beliefs are clearly irreconcilable with evolutionary ideas. According to a 2004 Gallup Poll, about 45% accept that God created humans in this way. Most Western religions, however, do not take a literal view of the Bible. This same poll indicated that 48% of the American public consider the Bible to be divinely inspired but not to be taken literally and 38% are theistic evolutionists, believing that biological evolution occurred over millions of years, with God guiding the process. Pope John Paul, for example, viewed a belief in evolution and a belief in God as perfectly compatible. Clearly, many citizens and religious leaders find it easy to accept evolution and to believe in God. They accept that evolutionary theory has its place: It explains adaptive features and the similarities and differences among organisms in the living world, but it does not tell us how to behave. It is not a theory of morality. That is the province of family, culture, and religion.

Creationist reasoners are likely to cite God or intelligent design as directly implicated in the origin of organisms. This is an intentional mode of reasoning in which animals are created to serve God's purpose. As long as science learners describe the evolutionary concept accurately, they should be assessed appropriately, even if they also express creationist ideas. Here are two examples of a creationist reasoner's response:

I would just explain it as God being the creator with infinite wisdom, and he designed and created every organism, down to the most minute detail....

I think they each were created as they are, with their own unique set of chromosomes, so I wouldn't have an answer how they would evolve.

Mixed Reasoners

Perhaps one of the more surprising findings from recent research is that many people do not consistently use a single pattern of reasoning. In fact, they are more likely to be mixed reasoners, employing two or three patterns of reasoning simultaneously. For some youth or adults this kind of reasoning may mark an unconscious transition between evolutionist and creationist reasoning or novice and informed naturalistic reasoning. In other cases, the conflict is more conscious, as it was for the parent who said, *I don't know what to believe, I just want my kid to go to heaven.* As described earlier, many youth

and adults propose evolutionary explanations for the origins of most animals, with the exception of humans (and sometimes other mammals as well). Here is an example from Activity 6 on human evolution:

I don't believe that they [humans] do evolve, because I don't believe necessarily in evolution. I mean yes I believe there's a Darwinism where the stronger species survived [like in the finches], but, I'm Christian so I believe God created man and God created chimpanzees....

In another kind of mixed reasoning pattern, we find creationist reasoning combined with informed naturalistic reasoning. The following response begins with creationist reasoning and concludes with a selectionist concept, as the science learner tries to explain why there are more large-beaked finches than small-beaked finches. It seems probable that the learner was unaware that she is describing an evolutionary mechanism, natural selection. This is a response from Activity 5 on finch evolution:

But like I said, I don't believe in evolution. So I don't believe that they evolved because it takes too long. There are too many failures before they evolve into something that finally works, so I just reject that view. Um, my guess would be that there probably were larger-beaked finches but there weren't as many of them and the small-beaked ones would have died out because they couldn't get the food.

You should not expect creationist reasoners to replace their belief system with Darwinian evolutionary concepts. But you can expect all of your science learners to understand evolutionary theory, so that they know how the natural world works. Many of them will be entering the kind of careers in which they will need this kind of understanding. How might they do this? By reflecting on their reasoning processes, they should be able to construct dual frameworks. You will have succeeded when you and they recognize that they are shifting from novice naturalistic or creationist reasoning patterns to informed naturalistic reasoning patterns, even if they use mixed patterns of reasoning.

Summary

Understanding how youth of different ages think and reason about evolution can lead to more effective guidance and teaching. By recognizing the nature of their thinking processes as they begin to acquire these complex and counterintuitive ideas, you can help them become better learners. The central idea is to have science learners engage in "intentional" conceptual change, in which learners are active agents in the learning process (Sinatra and Pintrich 2003). In this kind of conceptual change, learners reflect on the content of the learning process itself. Such a practice is more likely to result in lasting changes and will be an important mental tool for use in their daily lives. Teachers and youth leaders who are aware of their science learners' conceptual difficulties can use the ideas in this chapter in a variety of ways to suit the particular learning challenges that they face. These ideas represent some of the core concepts in inquiry-based and experiential learning.

References

American Association for the Advancement of Science (AAAS). 2001. *Atlas of science literacy.* Washington, DC: Author.

Bishop, B. A., and C. W. Anderson. 1990. Student conceptions of natural selection and its role in evolution. *Journal of Research in Science Teaching* 27: 415–427.

Evans, E. M. 2000. The emergence of beliefs about the origins of species in school-age children. *Merrill-Palmer Quarterly* 46: 221–254.

Evans, E. M. 2001. Cognitive and contextual factors in the emergence of diverse belief systems: Creation versus evolution. *Cognitive Psychology* 42: 217–266.

Evans, E. M., A. Spiegel, W. Gram, B. F. Frazier, S. Cover, and M. Tare. 2005. *Museum visitors explain seven evolutionary problems.* Evaluation report for the Explore Evolution project.

Hoagland, M. 1990. Reflections on the scientific process. *Dartmouth Medicine* Winter: 20–52.

National Academy of Sciences (NAS). 1998. *Teaching about evolution and the nature of science.* Washington, DC: National Academy Press.

National Research Council (NRC). 1996. *National science education standards.* Washington, DC: National Academy Press.

National Research Council (NRC). 2000. *Inquiry and the national science education standards: A guide for teaching and learning.* Washington, DC: National Academy Press.

National Science Teachers Association (NSTA). 2003. NSTA position statement on the teaching of evolution. In *NSTA Handbook 2004–2005.* Arlington, VA: Author.

Poling, D. A., and E. M. Evans. 2004. Are dinosaurs the rule or the exception? Developing concepts of death and extinction. *Cognitive Development* 19: 363–383.

Rosengren, K. S., S. A. Gelman, C. W. Kalish, and M. McCormick. 1991. As time goes by: Children's early understanding of growth in animals. *Child Development* 62: 1302–1320.

Sinatra, G. M., and P. R. Pintrich. 2003. *Intentional conceptual change.* Mahwah, NJ: Erlbaum.

Wellman, H. M., and S. A. Gelman. 1998. Knowledge acquisition in foundational domains. In *Handbook of child psychology: Vol. 2.* 5th ed., eds. W. Damon, D. Kuhn, and R. Siegler, 523–574. New York: Wiley.

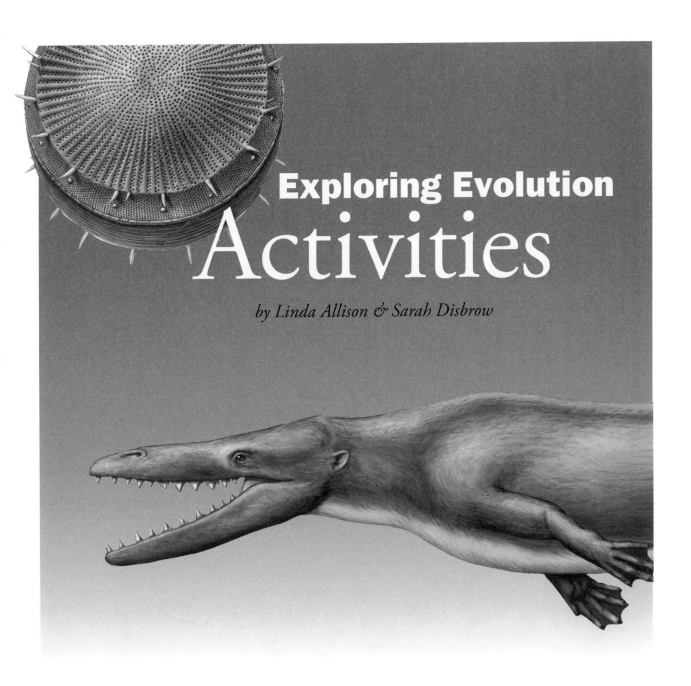

Exploring Evolution
Activities

by Linda Allison & Sarah Disbrow

UNSM Angie Fox illustration

Rodhocetus, the "barely walking whale." John Klausmeyer illustration.

Introduction
for Kids

UNSM Angie Fox illustration

At first glance, you may think a whale has nothing in common with a virus, but they actually have a great deal in common. They use the same mechanisms to make proteins. They also are both the product of four billion years of evolution.

This book aims to show you how scientists study evolution in some very different creatures. Evolutionary biologists try to understand how the fundamental principles of evolution have produced the diversity of life that surrounds us today.

Follow seven scientists to their labs and into the field as they make discoveries about how evolution works in creatures ranging from a small virus to a large whale. Each research project featured in this book highlights an important aspect of evolutionary biology, from the "arms race" between viruses and their human hosts to the long-term evolutionary changes that can turn a land mammal into a whale.

But how can scientists accept that evolution happens when nobody was around millions of years ago to watch it take place? It is important to remember that scientists are not always direct eyewitnesses to the events they seek to explain. No physicists have ever seen an electron; still they are confident they exist, because they can determine the effects of electrons through experiments. No geologist has ever seen lava rise from miles below the Earth's surface before it erupts out of a volcano's crater. But geologists can infer the origin of lava by looking at different lines of evidence such as the chemistry of lava rocks, and studying recordings of seismic activity around a volcano. Scientists study evolution in the same way, by looking at the evidence that's available, whether it is a fossil in the ground or bacteria in a lab.

In these activities you will investigate the theory of

evolution the way scientists do, through observations, hypotheses, and experiments. Scientists don't use the word *theory* in the informal way most people do, meaning a guess or a hunch. A scientific theory is an explanation for some aspect of the natural world, which makes sense of what would otherwise be a mysterious collection of data. The more a scientific theory holds up to the scrutiny of being tested through experiments or observations, the more it becomes accepted. Modern science is full of theories, from the theory of relativity to the germ theory of disease to the theory of evolution by natural selection. All of these theories have withstood many challenges from scientists and have come to be accepted as true by the scientific community.

Scientists have gathered a vast amount of evidence showing that life has evolved over the past four billion years, and that our own species is a product of that evolution along with the estimated 10 to 100 million other species with which we share the planet. The theory of evolution is so well-established that it has become the foundation on which modern biology is built. You can see its importance by reading the newspaper, listening to the radio or watching television news. A new strain of flu emerges, a new kind of oil-eating bacteria is discovered, resistance to antibiotics occurs in bacteria—these events show evolution in action.

Together, the activities in this book demonstrate the common principles of evolution at work in all life forms. Whether they are viruses, people, or whales, evolution occurs in all of them through the interaction of four basic processes: variation, inheritance, selection and time (or VIST for short). All living organisms show variation in their characteristics, such as color, size, and shape. Future generations inherit some of these characteristics through their DNA. Through interaction with the environment, these inherited characteristics undergo selection—some are better suited to the environment, and those are more likely to be passed on. Over time, the selected characteristics become more common in the population. Sometimes this can result in the formation of new species. Together, these four principles describe how evolution occurs.

Get ready to meet the scientists and their subjects: Go inside the body to meet a virus, underground to meet leaf-cutter ants, to the deserts of Pakistan to say hello to skeletons of ancient whales…and more. Try out the kinds of creative thinking scientists use to make new discoveries…and investigate evolution for yourself.

HIV:
Evolving Menace

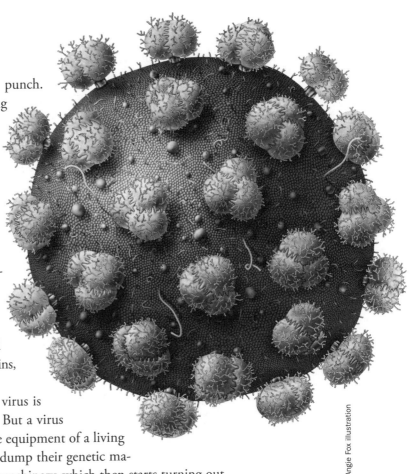

UNSM Angie Fox Illustration

Viruses are tiny, but they pack a big punch. Viruses spread and multiply fast, causing some of the most common and contagious diseases in the world. If you've ever had a rash, flu, or warts, you've probably hosted a few viruses. Actually, you've hosted a few billion. Once a virus sets up residence in an organism, it doesn't stay solo for long. One type of virus, HIV, can make 10 billion new viruses in a single day.

A virus consists of a small amount of genetic material inside a protein case. The genetic material of some viruses, such as herpes viruses, is DNA, while the genetic material of other viruses, like HIV, is RNA. DNA and RNA store information for making proteins, which in turn build a complete organism.

Whether it contains DNA or RNA, a virus is basically a recipe for making more viruses. But a virus can't make new copies by itself; it needs the equipment of a living cell. Viruses attach themselves to cells and dump their genetic material inside. The virus takes over the cell's machinery, which then starts turning out copies of viruses instead of its own products. The new viruses bud from the cell and go on to infect other cells. To viruses, your cells are nothing more than giant copy machines for making more viruses.

Luckily, your body is equipped with an infection-fighting army called the immune system. The immune system is made up of many different kinds of cells. There are T cells that alert the body to viruses and organize an attack, and other T cells that kill off the infected cells. There are B cells that produce proteins—called antibodies—that lock on to the virus and prevent it from attaching to a cell. And there are cells that clean up the mess. Your immune army is able to recognize and respond to virus invasions almost as fast as the viruses multiply. For example, it takes only a few days to recover from colds or flu. It happens naturally.

Some viruses, however, are so dangerous or deadly it isn't safe to let the immune system attack them naturally. You've probably received vaccines (usually shots) to protect you against highly contagious childhood diseases like polio, measles, mumps, and whooping cough. A vaccine is made of dead or weakened viruses. It boosts the immune system by stimulating your cells to create antibodies that can recognize and block the viruses before they attach to a cell. The vaccine prepares your body to recognize the infection if it should recur. In other words, your body becomes immunized against that particular virus.

But not all viruses are "vaccine friendly." For example, there are more than 200 types of cold viruses, too many for a single vaccine to handle. Other viruses, such as influenza, quickly evolve into new versions that may be resistant to last year's vaccine. Every year scientists need to predict the type of flu that will be widespread that year and develop a flu vaccine that targets that particular version. And you're likely to need a different vaccine the following year.

One of the deadliest viruses to emerge in recent years is HIV. It causes the disease called AIDS. HIV stands for Human Immunodeficiency (im-YOU-no-Dee-FISH-in-see) Virus. HIV invades the body's immune system, the very system that protects you from viruses and other invaders. Specifically, HIV targets the immune system's T cells. T cells organize the immune system response team. HIV is able to attach to a T cell through the interaction of its surface proteins with the T cell's receptor, like a key in a lock. Once the virus attaches to the cell, it fuses with the cell to deliver its genes inside.

People whose T cells become infected with HIV may not know it at first, but their immune system is gradually being destroyed, leaving them defenseless against deadly diseases such as pneumonia and tuberculosis. Even normally mild diseases like colds and flu may become deadly as the immune system collapses from HIV infection. This weakened condition is called AIDS (Acquired Immune Deficiency Syndrome). AIDS is always fatal. But with the proper drugs, people can live for many years with HIV infection.

HIV was discovered in the 1980s. In 20 years, the virus has spread to people in every country in the world. Today, about 40 million adults and children are infected with HIV, over 20 million have died from the disease, and the numbers keep rising. Increasingly, the victims are young women in their childbearing years. They often pass the virus to their babies.

As HIV races around the world, scientists are racing to stop it. One of these scientists is Charles Wood, a virologist (virus expert) at the University of Nebraska. Wood heads a team of scientists that is tracking the evolution of HIV in mothers and their infants. The team investigates how the virus evolves as it passes from mother to child, and how it changes in response to the immune system. Charles Wood studies

Charles Wood is the director of the Nebraska Center for Virology at the University of Nebraska. He studies how HIV is transmittted from mothers to infants.

Photo courtesy David Fitzgibbon University of Nebraska-Lincoln

these things to try to understand effective ways to block HIV's spread. It's not clear yet who will win this race, HIV or us. Read on and find out why.

HIV not only multiplies rapidly, it also mutates rapidly. This means that new viruses are not exact copies. They have genetic variations, or mutations. Scientists call these different viruses "strains." The strains are like a swarm of bees, each slightly different. The body's immune system (or a drug treatment) blocks many of the new strains and greatly reduces the number of viruses. But some of the viruses escape and produce new strains. Each time the immune system or treatment tries to elimi-nate the viruses, some of the new viruses escape, and the different strains build up into another swarm. This is typical behavior for viruses. What is unusual about HIV is that it infects the immune cells themselves. With each new swarm, HIV is simultaneously destroying the immune system, leaving the body defenseless against all kinds of disease.

Charles Wood's research takes him to a laboratory in Zambia, a country in Africa. HIV is one of the greatest threats to women and children in Zambia and in the rest of the developing world. In tracking the evolution of HIV from mother to child, Wood first identifies the strains of HIV in the mother on the day she has her baby. He then looks at what happens when the virus travels to its new host, the newborn baby.

To stay ahead of the evolving virus, scientists need to predict which parts of the virus are likely to change and evade recognition by the body's immune system, and which parts are likely to remain unchanged. Wood is hopeful: "If we can understand the evolution of HIV, where the virus is going, and why it is going there, we'll win." Wood and his team are collaborating with other scientists on a vaccine to give HIV-infected mothers before they give birth. The vaccine is designed to create antibodies that tie up all the viruses so that none are able to infect the baby.

But what about a vaccine for the baby? Try your hand at creating one that works. In this activity you'll meet the virus, HIV, investigate what happens when it infects a newborn, work on a vaccine to save the baby, and discuss the evolutionary race between humans and HIV.

Zambia is in central Africa.
SMM Lonnie Broden illustration

Women and children with AIDS-awareness pins at a clinic in Zambia.
Photo courtesy John West

PART ONE
Inside the invader

A virus has been called "a piece of bad news wrapped in a protein." Meet HIV, then make a flip book movie to watch the action of a virus attack.

Work with a partner

Each team of two will need:
- Virus Flip Book sheets 1, 2, and 3
- clip or stapler
- marker
- colored pencils
- scissors

1 Map a Virus

A virus is essentially a recipe for making itself, packaged in a protein overcoat. HIV targets one kind of human immune cell, called T cells, and then breaks in. Once inside it pirates the cell's control center, takes over the cell and makes thousands of copies of itself. Check out the virus parts. Draw lines connecting the labels to the features on the HIV diagram:

a Find the virus envelope proteins. They are the spikes or keys on the surface that the virus uses to break into living T cells.

b Find the long thin strand of RNA located inside the virus. RNA is a genetic code for making more of this type of virus.

c Find the enzymes. They are chemicals (represented by dots) that help the virus reproduce.

HIV (a virus)

Envelope
Proteins

RNA

Enzymes

2 Explore a T Cell

T cells are one kind of the body's defense cells. They respond to an infection by identifying the invader and signaling the rest of the body's immune system to get busy defending against the attack.

Draw lines connecting the labels to the features on the T cell diagram:

a Find the nucleus. This is the control center of the cell.

b Inside the nucleus, find the strands of DNA. This is the genetic material that contains the recipe for making a new cell.

c Find the receptors on the cell membrane. These are the locks that the virus must open to gain entrance into the T cell.

T cell (an immune cell)

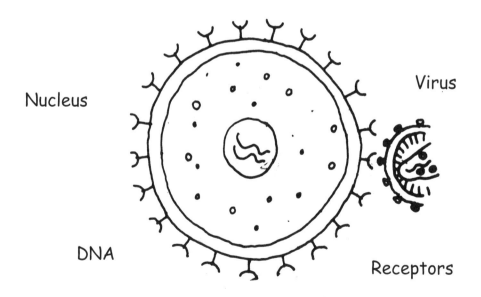

Nucleus

Virus

DNA

Receptors

3 What's the Difference?

How is a virus similar to and different from a T cell?

4 Make a Flip Book Movie: Virus Attacks

a First cut out all the flip book pages. The movie will work better if you are exact in your cutting. Put the pages in order, starting with page one.

b Color the pages if you like. Be sure to keep the colors the same on each page.

c Line up the outside edges without the numbers by tapping them on a flat surface.

d Test flip your movie. When you are sure your pages are in good flipping order, bind the pages with a staple, clip, or rubber band. Be sure to bind the flip book pages on the side with the number.

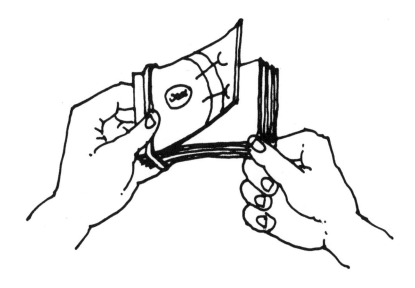

Virus and the Whale: Exploring Evolution in Creatures Small and Large

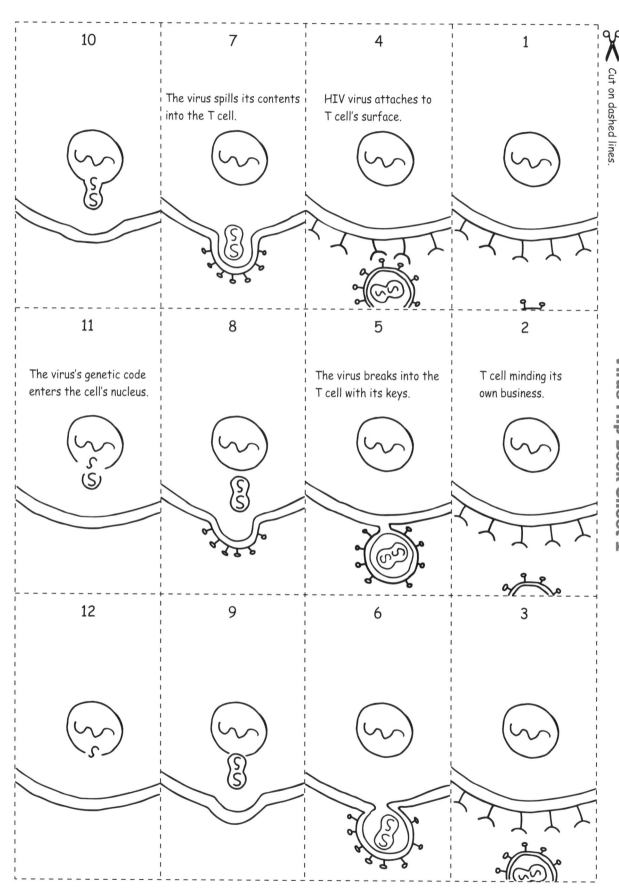

Virus Flip Book Sheet 1

10	7	4	1
	The virus spills its contents into the T cell.	HIV virus attaches to T cell's surface.	

11	8	5	2
The virus's genetic code enters the cell's nucleus.		The virus breaks into the T cell with its keys.	T cell minding its own business.

12	9	6	3

HIV: Evolving Menace

22

New virus forms
inside the T cell.

19

16

New virus code directs
the T cell to make virus.

13

23

20

17

14

The virus splices into
the T cell's DNA.

24

21

18

15

The virus inserts its genetic
code into the T cell's DNA.

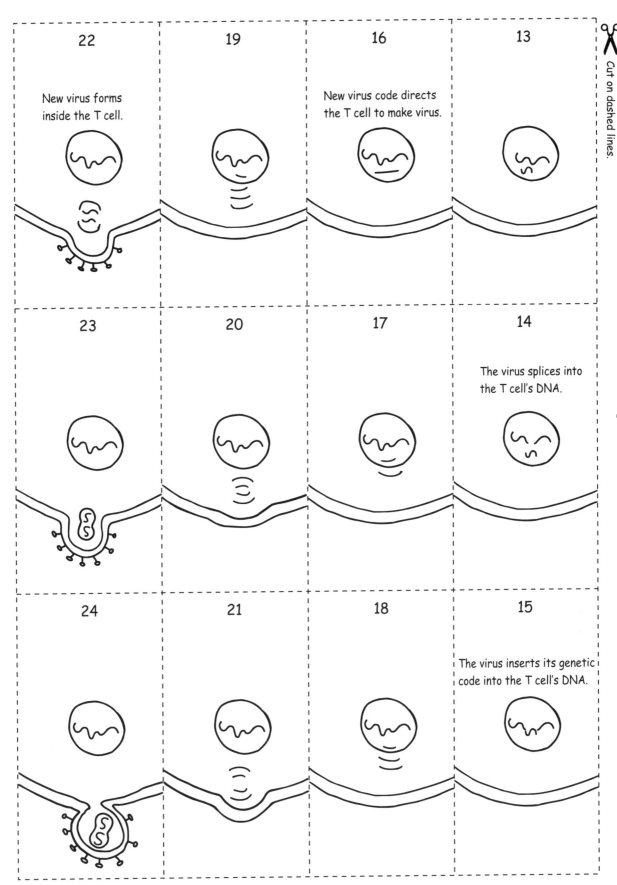

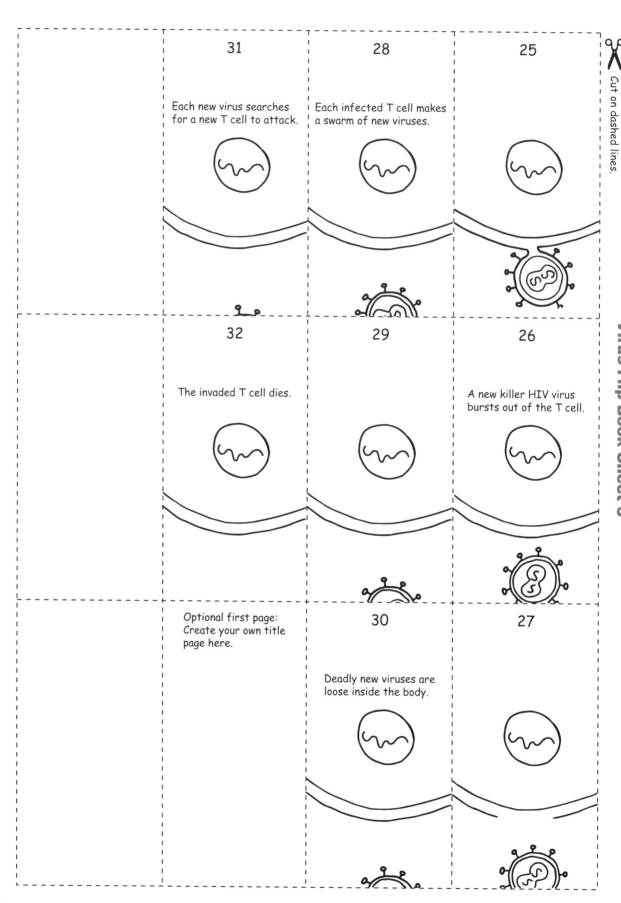

31

Each new virus searches for a new T cell to attack.

28

Each infected T cell makes a swarm of new viruses.

25

32

The invaded T cell dies.

29

26

A new killer HIV virus bursts out of the T cell.

Optional first page: Create your own title page here.

30

Deadly new viruses are loose inside the body.

27

e   Read the script of what is happening in the movie on the pages. Use these as clues to caption the movie characters in the next step.

f   The characters in your movie need labels. Write the following label names on the flip book pages where they first appear:

• T cell

• T cell nucleus

• T cell genetic code

• HIV

• HIV genetic code

• New virus

g   Write a title for this movie. Put it on the first page of the flip book.

5 Consider This

If you had to write an epilogue (comment about what will happen next) for this script, what would it say?

PART TWO
Evolution of the mutants

In this section you will investigate how HIV from a mother can evolve into new strains when it is passed on to her baby. You will look closely at a section of the virus RNA code as it changes over generations.

DNA and RNA store information for making proteins, which in turn build a complete organism. Your challenge is to create a "vaccine" that can disable all the virus strains and save the baby's life. In this activity the vaccine will work on the viral RNA. In real life, vaccines work directly on the proteins made by the virus.

Work with a partner
Each team of two will need:

- Virus Mutation Tracker sheet
- Available "Vaccine" (see page 58)
- Your "Vaccine" (see page 58)
- scissors
- coin
- colored pencils: yellow, red, blue, green

1 Tracking Mutations
Take a look at the Virus Mutation Tracker sheet. The columns on the sheet represent a short segment of the HIV RNA code. A real virus has about 10,000 sites. You will work with only 10 sites of code.

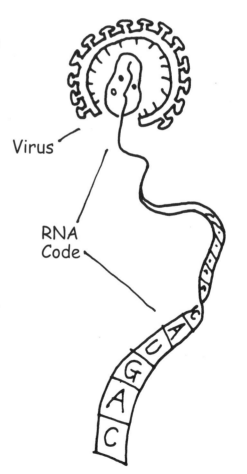

Virus

RNA Code

VIRUS MUTATION TRACKER

	Mother's Virus Strain	Infant Virus Strain 1	Infant Virus Strain 2	Infant Virus Strain 3	Infant Virus Strain 4	Infant Virus Strain 5
Site 1	A	A	A	A	A	A
Site 2	U	U	U	U	U	U
Site 3	U	U				
Site 4	G	G	G	G	G	G
Site 5	C	C	C	C	C	C
Site 6	C	C				
Site 7	A	A	A	A	A	A
Site 8	C	C	C	C	C	C
Site 9	G	G				
Site 10	C	C	C	C	C	C

Color Key A=red/Adenine U=blue/Uracil C=yellow/Cytosine G=green/Guanine

2 Create a Virus Strain

a   On the Virus Mutation Tracker sheet, notice the column on the left called Mother's Virus Strain. It is made up of a string of letters: A-U-C-G. Those letters represent the nucleotides Adenine, Uracil, Cytosine, and Guanine.

Why are we using Uracil? The DNA in humans and some viruses uses four nucleotides as building blocks: Adenine, Thymine, Cytosine, and Guanine. But HIV belongs to a group of viruses that has RNA instead of DNA. RNA uses building blocks: Adenine, Uracil, Cytosine, and Guanine.

b   Color each site of the Mother's Virus Strain according to this chart:

Nucleotide Color Chart

LETTER CODE	COLOR IT	NUCLEOTIDE
A	Red	Adenine
U	Blue	Uracil
C	Yellow	Cytosine
G	Green	Guanine

c   Now color each site of the next column, Infant Virus Strain 1, according to the Nucleotide Color Chart. How does it compare to the mother's strain?

When this infant was born the mother infected her infant with her virus, Strain 1.

3 Create a Mutant Strain

a   Inside the baby, the virus invades cells and reproduces. HIV often makes mistakes when it copies itself. Instead of a perfect copy, certain sites in the code tend to switch nucleotides.

b   To create Infant Virus Strain 2, find the third column from the left, Infant Virus Strain 2. First color each site of Strain 2 according to the Nucleotide Color Chart, except for sites 3, 6, and 9. These sites switch letters at random. Use the information in the Coin Toss Chart to fill in the random letter.

Coin Toss Chart

COIN TOSS	LETTER CODE	COLOR IT
Heads/Heads	A	Red
Heads/Tails	U	Blue
Tails/Heads	C	Yellow
Tails/Tails	G	Green

c Flip a coin two times to decide if an A, U, C, or G will fill site 3. Then flip twice again for site 6, and again for site 9. Then finish coloring the code. Congratulations, your virus has replicated!

Coin to toss

Tracker sheet

Colored pencils

4 Make More Mutants

To make more virus strains, repeat the above process three more times to create Strains 3, 4, and 5. The infant now has five different strains of viruses replicating inside its body. Your next challenge is to make a vaccine before the baby's immune system is overwhelmed by virus attacks.

5 Discover a Vaccine

Your infant is now host to a number of different virus strains having different genetic codes (sequences of A,U,C,G). Your job is to discover a "vaccine" with a three-letter sequence that can kill all the infant's viruses.

a A "vaccine" is available that targets Infant Virus Strain 1. It will disable any virus that has the three-letter sequence, G,C,C (reading from top down). Color the "vaccine" according to the Nucleotide Color Chart (A=red, U=Blue, C=Yellow, and G=Green). Then cut it out.

b Use the GCC "vaccine" to scan the other virus strains. If the GCC sequence matches, it will disable that strain. The infant will be saved when all the strains are disabled. How many strains is this vaccine effective against?

c Try creating your own three-letter vaccine with any three-letter sequence of A,U,C,G's. Use your new "vaccine" to scan all of the virus strains. How many strains is your new vaccine effective against?

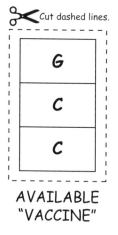

Cut dashed lines.

AVAILABLE
"VACCINE"

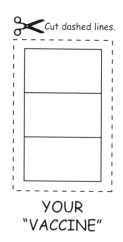

Cut dashed lines.

YOUR
"VACCINE"

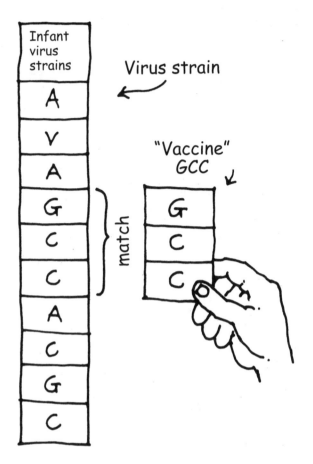

6 Consider This

Is it possible to create a vaccine that disables all the infant's virus strains?
Why or why not?

PART THREE
HIV, information please

It takes a lot of effort, money, and time to develop a drug for HIV. Because the virus evolves so quickly, constantly creating new strains, no drug is effective for long. In North America, drugs that help someone stay alive with HIV cost thousands of dollars a year. In Africa, until recently, an average person in a country like Zambia could not afford these drugs. Getting people information about how HIV works and how it spreads is one cheap and effective way to help prevent this deadly disease. But even after decades of research, there is still confusion and fear about HIV and AIDS. Read the HIV fact cards. Then turn some of the information that you learned into a poster to inform others.

Work with a partner

Each team of two will need:
- HIV Fact Cards 1 and 2
- poster-sized paper, collage materials, or magazines
- glue stick
- markers or colored pencils
- scissors

1 HIV Cards: just the facts

a Cut the HIV Fact Cards apart.

b Deal the cards out equally between you and your partner.

c Read the cards. Report back to the group a summary of what you read.

2 Display your Knowledge: create a poster

a Pick one fact from the cards that you think is the most useful or most interesting to know. Use that fact to make an HIV Information Poster.

b First design the poster's text. (Short and simple is best.)

c Then create an illustration using a drawing or collage. Share your poster ideas with the other groups.

3 Consider This

What do you think is the most important thing your friends or family need to know about HIV and AIDS?

HIV Fact Cards 1

✂ Cut on dashed lines.

HIV TAKES OFF

Scientists suspect that HIV may have first emerged in the 1930s and for decades was restricted to central Africa.

Gradually people in the region became more mobile, moving to growing cities, spreading the virus. Increasing air traffic allowed the virus to reach other continents. By 1980 it was rapidly spreading in Europe, the Caribbean, and North America.

HIV vs. AIDS

People who get infected with HIV virus may not know that they are infected for years. They feel fine and look no different than before.

The only way to know if people have HIV at this stage is to give them a special blood test. But after several years, most people who carry HIV become very sick. At this stage the condition is called AIDS (Acquired Immunodeficiency Syndrome).

HIV HISTORY

Scientists believe that HIV first emerged in Africa in the early 1900s. It evolved from viruses that infect chimpanzees and monkeys. As hunters killed these primates for food, they sometimes became infected with the virus. Over time, they began to spread the virus to other people.

VERY VARIED VIRUSES

The variation found in viruses such as HIV is enormous compared to the variation found in animals or plants. One scientist describes them this way: The number of potential variants of HIV in one person per day can be as many as 1,000,000,000.

AIDS APPEARS

Doctors first recognized AIDS in the early 1980s when they noticed a growing number of patients who were dying from diseases that rarely harmed healthy people. It was not until 1983 that French scientists isolated HIV, the virus that causes AIDS.

HIV: YOU CAN'T GET IT BY...

HIV virus travels in body fluids of infected people.
You can't get it from:
- Hugging
- Shaking hands
- Just being close
- Coughs or sneezes
- Sweat
- Mosquito bites
- Towels, telephones, swimming pools

Linda Allison illustration.

HIV Fact Cards 2

HIV: HOW IT SPREADS

People can only become infected with HIV if they come into direct contact with body fluids of an infected person. This can happen through sexual contact, contaminated blood transfusions, or sharing needles. Pregnant women can pass HIV to their unborn children, because their bloodstreams are connected. HIV can be passed to a baby through mother's milk.

FINDING A CURE

HIV stands for human immunodeficiency virus. Currently, there is no effective vaccine for HIV. It is possible, however, to keep the virus in check with a combination of drugs. If people with HIV stop taking the drugs, they will get sick again.

HIV's KILLER RESISTANCE

As HIV spreads, it also evolves. The virus has experienced natural selection as it has been attacked by the immune systems of its human hosts. Over thirty drugs and treatments have been developed to fight HIV. In every case the virus has evolved resistance.

HIV FIGHTERS

A series of HIV-fighting drugs has been introduced since 1987. The most effective treatment so far is a combination of drugs that interfere with the virus's ability to copy itself.

The drugs are very expensive to produce and to dispense. The vast majority of people in the world infected with HIV cannot afford to take them.

HIV DEATHS

More than 20 million people worldwide have died of AIDS since 1981. AIDS experts worry that the toll will climb even faster if new epidemics break out in countries such as China and India.

HIV ATTACK

HIV (human immunodeficiency virus) gradually destroys the immune system. Infected people may feel and look fine for many years while this happens. Eventually their immune systems collapse, and they fall victim to other diseases like pneumonia. At this point the condition is called AIDS (Acquired Immune Deficiency Syndrome).

HIV: Evolving Menace

PART FOUR
Be a science reporter

Write a short news story about HIV. Tell your readers about how a baby who is born with HIV carries exactly the same strains of the virus as the mother. Within six months, the baby can have millions of new strains. Based on what you have learned, explain how you think the baby gets new strains of HIV.

P.S. Don't forget the headline.

DIATOMS:
One-Celled
Wonders

Diatoms (DIE-a-toms) are one of the most important things you never knew about. They are everywhere there is water. A drop of lake water is packed with them. You probably swallow millions every time you go swimming. These tiny, one-celled life forms populate the world's ponds, rivers, and oceans (and anywhere else that's the least bit wet). They spend their invisible lives quietly using sunlight to turn carbon dioxide and water into food and oxygen. Diatoms are the basis of the food chain, and they produce much of the oxygen you breathe.

Dead or alive, diatoms are important. Their skeletons are made from a substance called silica, which is the main ingredient in glass. When diatoms die, their dead bodies pile up by the trillions on the floor of oceans and lakes, forming a chalky mud. This plentiful mud is used as a water filter, an insecticide, and even as an ingredient in some toothpaste.

More important still, dead diatoms fossilize (become preserved) over time. The fossil record can tell us many things about Earth's past, and diatoms are especially good record keepers. First, their fossils are plentiful. A dollop of mud the size of an ice cube may contain as many as 50 million well-preserved diatoms. Second, there are many different kinds of diatoms. No one knows the exact number, but scientists estimate a quarter of a million species, each one preserving a unique record of the past. And third, diatoms are so sensitive to changes in temperature and moisture that the presence or absence of a particular species can tell us a great many things about the environment. From the rise and fall of different diatom populations, we can learn about changes in past climates, which in turn can give us more clues about the world that plants and animals occupied in the past.

This is where four scientists, a famous lake, and a mysterious diatom enter our story. Brian Shero, a professor at a small New York college, was studying diatoms in Yellowstone Lake when he found a diatom he had never seen before. He sent it to a diatom expert, Edward Theriot, to identify. Ed's life goal has been to learn everything he can about diatoms. He was surprised and excited to discover a diatom he didn't know. It was a species that was brand new to science, and more amazing still, there was no record of it living anywhere else on Earth. In honor of its hometown lake, the new diatom was named *Stephanodiscus yellowstonensis* (Steff-ah-no-DISK-us yellow-stone-EN-sis), which means crowned disk of Yellowstone.

Ed was curious about the origins of the mysterious diatom. How and when did it appear in Yellowstone Lake? Did it arise from some other diatom species already there? To find out answers to these questions, Ed and a team of scientists searched for fossil diatoms in mud samples from the bottom of the lake. Yellowstone Lake was formed after the last Ice Age, around 14,000 years ago. Ever since then diatoms have lived and died in the lake. Their sinking skeletons have built up fossil deposits dozens of feet thick, creating a unique time line for scientists to study.

The name of the diatom *Stephanodiscus yellowstonensis* means "crown disk of Yellowstone."
UNSM Angie Fox illustration

Yellowstone Lake is located in Yellowstone Park.
SMM Lonnie Broden illustration

Diatoms: One-Celled Wonders

Yellowstone Lake in Yellowstone National Park contains a species of diatom that is found nowhere else.
Photo courtesy Joe Sawyer

Scientists take a core sample from beneath Yellowstone Lake.
Photo courtesy Edward Theriot

To actually see the fossil deposits is difficult work. The scientists drove a tube a few inches wide and as long as a telephone pole down through hundreds of feet of water and lake-bottom sediments. Sediments in a lake settle gently year by year, forming undisturbed layers often as distinct as the layers of a birthday cake. When the long tube was pulled up, it revealed layers of mud in clear stripes. Each stripe represented a year of diatoms. "We can look in there and find the sediment from 11,298 years ago, in the summer," Ed says. The core sample covered the entire geological history of the lake, from its birth 14,000 years ago to the present.

What Ed found in the core sample blew him away. By identifying and counting diatoms sample by sample, he could see evolution in action. When Yellowstone Lake first formed, a diatom lived there called *Stephanodiscus niagarae* (Ny-ah-gary), the Niagara diatom. There was no sign yet of *S. yellowstonensis*, the Yellowstone diatom. Over the next 4,000 years, some Niagara diatoms began to change, resembling *S. yellowstonensis* more and more. "We can watch, literally watch, as these diatoms go from one species to another," Ed exclaimed. In the core sample, the scientists were able to find

The diatom *Stephanodiscus yellowstonensis* (right) has about five ribs between each spine. Counting the ratio of spines to ribs is the way diatom experts tell this species from similar ones.
UNSM Angie Fox illustration

Edward Theriot (below) is a diatom scientist at the University of Texas at Austin.
Photo courtesy Marsha Miller

Sheri Fritz (bottom) studies how climate change influences the evolution of diatoms.
Photo courtesy Tom Simons University of Nebraska-Lincoln.

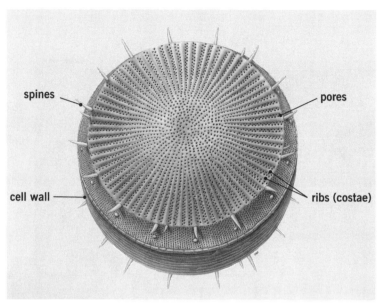

a diatom that had some parts that looked like the Yellowstone diatom and some parts that looked like the Niagara diatom. When a form shares features with two different species, it is sometimes called a "transitional form." Not only did the core sample show that a new species had evolved, but it also proved to be the fastest evolution so fully demonstrated in the fossil record.

But how did it happen? Ed wondered what else was changing while the diatom was developing new features. He teamed up with Sherilyn Fritz, a geologist who studies how lakes change through time. Sheri looked at the amount of volcanic ash in the layers and at other fossils. She sent mud to her colleague Cathy Whitlock to look for pollen, the tiny grains that flowering plants produce. Like diatoms, pollen grains fossilize, and they are helpful in tracking changes in plant life, temperature, and rainfall.

Ed and Sheri then analyzed the data and came up with an interesting theory about the mysterious appearance of the Yellowstone diatom. In this activity you will have a chance to explore the world of diatoms, meet the new diatom, *S. yellowstonensis*, learn how to make a core sample, check out Ed and Sheri's data, and see if there is a connection between diatoms and climate change.

PART ONE
Meet the diatoms

Diatoms are dazzlingly beautiful little organisms that live in oceans, lakes, rivers, and puddles. . . just about everywhere you find water. Explore the tiny world of diatoms with a deck of diatom cards.

Work in small groups or with a partner

Each team will need:
- Diatom Fact Cards 1–3 (cut, fold, and tape 12 diatom cards)
- sheet of paper
- tape
- scissors

1 Sizing Up Diatoms

Look at the diatom pictures on the cards. Diatoms come in many sizes. So if diatoms are everywhere, why haven't you ever seen one? Because they are small. How small?

Look at this circle: o. Guess how many diatoms it would take to fill in the circle. (The answer is on the last page of this activity).

2 Discovering Diatom Shapes

Explore how beautiful diatoms are to behold.

a Sort the diatom cards into groups or families in a way that makes sense to you. As you sort the cards, don't forget to read the general diatom facts on the back of the cards.

b What feature or features did your group use to sort?

c Share your sorting with the other groups. Did every group sort the same?

Diatom Fact Cards 1

Diatoms grow hard, translucent shells of silica that are very much like glass.

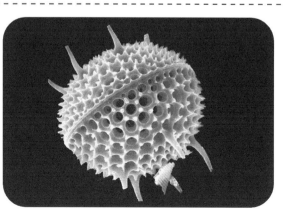

Diatoms live in both fresh and salt water, rivers, lakes, and ponds. They can be found almost everyplace on Earth that is wet.

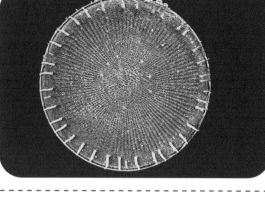

Diatoms reproduce by splitting their top and bottom halves apart and growing another half shell.

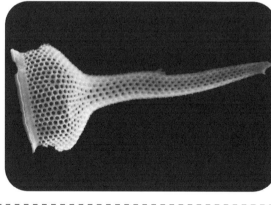

Diatoms are so sensitive to chemical changes in water that scientists use them as markers to track pollution.

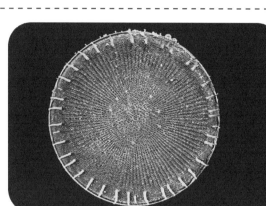

✄ Cut on dashed lines, fold on solid line.

Photos courtesy David Harwood and Vladimir Nikolaev.

Diatom Fact Cards 2

✂ Cut on dashed lines, fold on solid line.

Small diatoms measure 15-20 micrometers(twice the size of a red blood cell, too small to see). The biggest diatoms measure up to 200 millimeters (as wide as a pencil line).

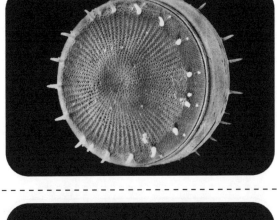

Diatomaceous earth (diatom powder) is used in pool filters. The tiny holes in diatom shells make them an excellent filter material.

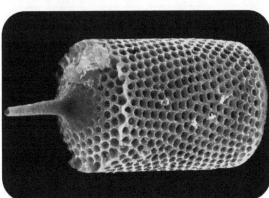

Diatoms have no equipment for moving effectively on their own. They drift about in water or slide on ooze at the bottom of lakes.

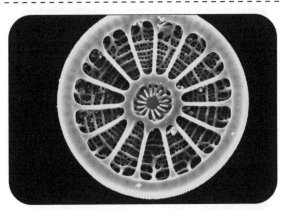

Like land plants, diatoms use sunlight to turn carbon dioxide and water into food and oxygen.

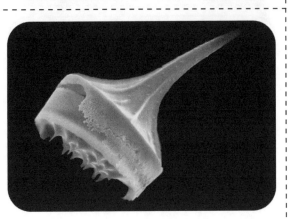

Photos courtesy David Harwood and Vladimir Nikolaev.

Diatom Fact Cards 3

Police sometimes use diatoms to help solve crimes. The unique mix of diatoms in a lake or river can pinpoint a water source. This has proved handy in identifying where a victim with lungs full of water originally died.

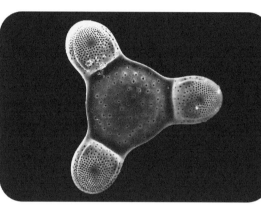

Diatoms are single-celled organisms. There are more than 250,000 species of diatoms. Each one has a unique shape.

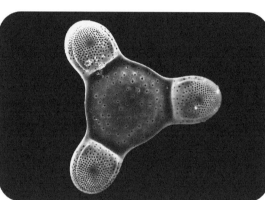

Pond scum is another name for diatoms. The brown slime on stones at the water's edge is very likely a huge population of tiny diatoms.

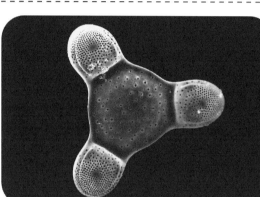

Probably the most important life form on Earth, diatoms are constantly at work turning sunlight into food. They are a basic food for many other animals from tiny shrimp to giant whales.

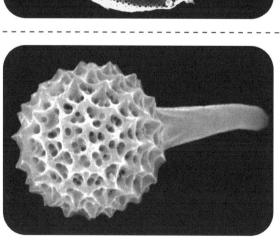

3 Fascinating Facts

Recall your favorite diatom fact. Write it in the form of a question. Use your question to quiz the rest of the group.

4 Who's Closely Related?

Choose two diatoms that you think are most closely related. Tape them in the boxes below. Why do you think they are closely related?

Discovering Diatom Shapes

Tape your card choices here

5 Diatom I.D.

The new diatom that Ed and Sheri studied was named *Stephanodiscus yellowstonensis* after the lake it was found in. It is very similar to another species, *Stephanodiscus niagarae,* which is found in neighboring lakes, but not in Yellowstone Lake. Both diatoms can be identified on these scanning electron microscope (SEM) images by their round shape like a bicycle wheel with many thin ridges or "ribs" radiating out from the center.

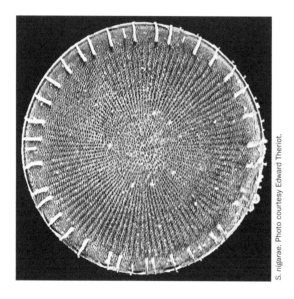

S. *nigarae.* Photo courtesy Edward Theriot.

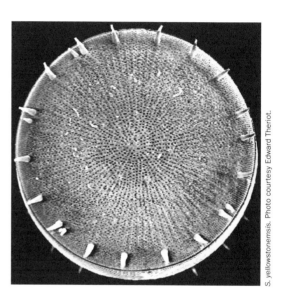

S. *yellowstonemsis.* Photo courtesy Edward Theriot.

6 Consider This

Look at the drawings of the two species of diatom shown below.

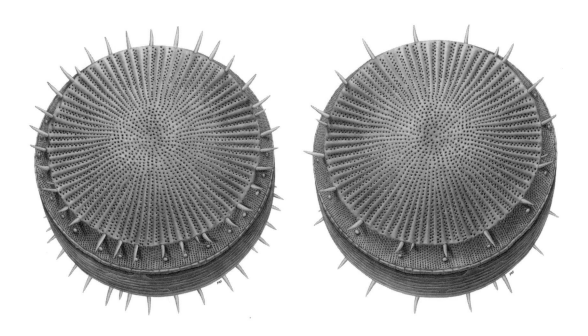

How do you know they are different species?

Hint: The experts count the number of "ribs" between the pointy "spines" jutting out from the ends. How many ribs are there for every spine in the Niagara diatom? How many ribs to spines are there in the Yellowstone diatom?

PART TWO
Sampling the past

To solve the mystery of where the Yellowstone diatom came from, two scientists teamed up to share their knowledge. Sheri Fritz, a geologist, took core samples (samples of mud collected in a tube) from the bottom of the lake. Ed Theriot, a diatom expert, helped identify ancient diatoms in the core samples. In sorting out the evidence, Sheri and Ed discovered that the Yellowstone diatom had a unique history. They believe this diatom provides the most complete fossil record of the evolution of a new species in the fossil record.

Check out Sheri and Ed's core sample findings. Then make your own muddy history with layers of "mud," and take a core sample. Then use a scientist's methods to reconstruct your own series of past muddy events.

Work with a partner
Each team of two will need:
- Your Pollen Chart
- Your Diatom Chart
- Read the Core Sample
- one small cup (5 oz, clear is better)
- one clear plastic straw
- brown sugar (9 teaspoons)
- three different bright colors of sugar (1/2 teaspoon each)
- paper plates (3)
- plastic spoon
- cm ruler
- newspapers (to cover work tables)

1 Tracking Diatoms
To begin their core sample, the scientists dug down through the mud at the lake bottom until they reached a layer that dated from 14,000 years ago. They removed the core sample and took it back to the lab. There they took tiny scoops of mud from the core every few centimeters. By carefully washing and sorting the contents, they were able to learn which diatoms lived in a specific layer.

a   Read the Diatom Chart. When did the Yellowstone diatom first show up?

Diatom Chart

Date of core sample	*Stephanodiscus* DIATOM VARIETY
Now	S. *yellowstonensis*
2,000 years ago	S. *yellowstonensis*
4,000 years ago	S. *yellowstonensis*
6,000 years ago	S. *yellowstonensis*
8,000 years ago	S. *yellowstonensis*
10,000 years ago	S. *yellowstonensis*
12,000 years ago	S. (transitional)
14,000 years ago	S. *niagarae*-like ancestor

b A form that shares features with two different species is sometimes called a "transitional form." When does the transitional form between the Yellowstone diatom and the Niagara diatom show up in the core sample?

c Look back at the previous activity section to see what differences you might expect to find between the Niagara diatom and the Yellowstone diatom. List them here.

2 Tracking Plant Pollen

If you have ever stuck your face in a flower and come away with a dusty nose, you know about pollen. Plants produce huge numbers of pollen grains to fertilize their seeds. These tiny grains settle everywhere, including in the mud at the bottom of lakes. Under magnification the tiny grains have a distinct shape unique to each species of plant. Pollen trapped in ancient mud tells what plants were alive when that mud layer was formed. Because different plants thrive in different climates, knowing the kind of plants alive at each time provides good clues to past climate conditions.

Here's how to make some sample "muds." Then drill your own core sample, and use your pollen finds to make sense of the past climate.

a Make different muds that contain pollen. The whole group should decide on which color sugars will represent which kind of pollen (e.g. green for sage and grass pollen, red for lodgepole pine pollen). Everyone should follow the same rules. Put 3 teaspoons brown sugar on a plate. Add 1/2 teaspoon of one color of the sugar that will represent sage and grass pollen. Stir and label this plate "sage and grass pollen." Lumps are fine.

b On another plate, put 3 teaspoons brown sugar and add 1/2 teaspoon of the color of sugar that represents lodgepole pine pollen. Label its plate.

c Put 3 teaspoons of brown sugar on a third plate. The brown sugar represents mud of the glacial period. The frozen climate permitted little plant growth and few pollen grains.

d Write the pollen (sugar) colors on Your Pollen Chart to help you identify your pollen types.

Your Pollen Chart

POLLEN COLOR (SUGAR COLOR)	TYPE OF POLLEN	HABITAT	CLIMATE
	Sage, grass	Meadows, grasslands	Wet, cool
	Lodgepole pine	Thick forest	Dry, warm
	None	Icy glaciers	Frozen, cold

e Diatoms are represented by the third color of sugar. Decide what kind of climate conditions your diatoms require. Check one type.

_____ Wet cool,   _____ Dry, warm   _____Frozen, cold

Add 1/2 teaspoon of the diatoms (third color of sugar) to that colored climate layer. Fill in the chart below.

Your Diatom Chart

YOUR DIATOM'S COLOR (SUGAR COLOR)	CLIMATE YOUR DIATOM REQUIRES

3 Make History

a Gather your materials with your partner. Put the cup on a newspaper. Choose a mud and put it in the cup. Make a layer at least one centimeter thick (more is fine).

b Add a second layer of different mud.

c Continue adding layers (up to 5 more) and record the order below.

"muds"

newspaper

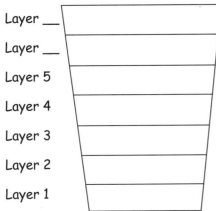

Layer __

Layer __

Layer 5

Layer 4

Layer 3

Layer 2

Layer 1

4 Take a Core Sample. Read the Layers

Your team will take a core sample of another team's cup. Your challenge is to map the pollen types in the layers to make a climate history and determine where the diatoms appear.

a Swap cups with another team. Take a core sample by pushing a straw straight down through the layers.

b Put your finger on the open end of the straw and remove the core by pulling up. (Keep your finger on the straw to keep the material inside.)

c Dust off the outside of the straw so you can see the layers. Read your core's history by recording each layer. Fill in the information for each layer starting from the bottom up, using both words and a drawing on the chart below.

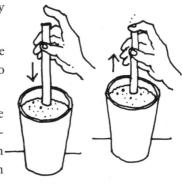

d What was the weather like in the oldest period in your core sample?

Read the Core Sample

DRAW CORE SAMPLE HERE	LAYER # Counted from the bottom up	TYPE OF POLLEN IN LAYER	CLIMATE TYPE	CHECK IF DIATOMS ARE PRESENT
	Layer 5			
	Layer 4			
	Layer 3			
	Layer 2			
	Layer 1			

5 Consider This

How did the climate change from the oldest layer (#1) to the most recent?

PART THREE
Mystery in Yellowstone Lake

Sampling pollen in ancient mud is a great tool for figuring out weather conditions of the past. Sheri's friend, Cathy Whitlock, analyzed the pollen found in the core samples from the mud at the bottom of Yellowstone Lake. The pollen gave clues about the climate conditions thousands of years ago. Now is your chance to analyze the scientists' data.

Work with a partner
Each team of two will need:
- Core Sample Chart
- Pollen Chart (from Part Two)

1 Climate Clues
Use the pollen data to create a 14,000-year weather report of Yellowstone Lake.

a Fill in the Core Sample Chart. Use the Pollen Chart as the key to identify the plant pollen found in the core sample. Be sure to fill in both pollen type and the climate conditions.

b Write a weather report for Yellowstone Lake for the past 14,000 years. Summarize the trends.

c When did the transitional form of the Yellowstone diatom begin to appear in the core sample? According to the pollen record, what else happened about this time?

Core Sample Chart

DATE OF CORE SAMPLE	DIATOM VARIETY	POLLEN TYPE	CLIMATE
Now	S. yellowstonensis	name:	
2,000 years ago	S. yellowstonensis	name:	
4,000 years ago	S. yellowstonensis	name:	
6,000 years ago	S. yellowstonensis	name:	
8,000 years ago	S. yellowstonensis	name:	
10,000 years ago	S. yellowstonensis	name:	
12,000 years ago	S. (transitional)	names:	
14,000 years ago	S. niagarae-like ancestor	names:	

2 Consider This

No one can say what exactly caused the Yellowstone diatom to evolve. We do know that diatoms are extremely sensitive to environmental conditions like available light, nutrients, and temperature changes. What changes do you think might have affected the evolution of this new diatom?

PART FOUR
Be a science reporter

Write a short news story about diatoms. Tell your readers about a new species of diatom that is found only in Yellowstone Lake. From core samples taken beneath the lake, scientists can tell that the Yellowstone diatom first appeared thousands of years ago, at a time when the climate was dramatically changing. Based on what you have learned, explain how you think this new diatom came to exist in the lake.

P.S. Don't forget the headline.

ANTS & CO.:
Tiny Farms

Can a farm be the size of a walnut? It can if the farmer is an ant. Some of the world's smallest farmers are leaf-cutter ants. Leaf-cutter ants farm in the tropical zones of the Western Hemisphere. Farming in the tropics is no easy achievement, whether you're an ant or a human, because the warm soils swarm with pests ready to attack your crop. These six-legged farmers have amazed scientists by growing crops successfully for millions of years. By comparison, humans have only been farming for a few thousand years. What is the secret of the ants' long-standing success? Cameron Currie, a biologist, discovered the answer by taking a close look at life on an ant farm.

Leaf-cutter ants harvest leaves from the tropical forest, but they can't eat these leaves. Most tropical leaves are poisonous to the ants. Instead, the ants feed the leaves to a fungus. The fungus eats the leaves and produces nourishing food that the ants can eat. The fungus is similar to mushrooms you buy in grocery stores.

Every day the ants file out of their nests and into the forest. They cut down piec-
es of leaves, hoist them over their bodies, and haul them back to their nest. The ants chew the leaves into a pulp and spread the pulp deep in the underground nest. The fungus grows on the bed of leaves, and the ants weed and tend it carefully, eating only the fungus tips. When a queen ant establishes a new nest, she brings along a piece of fungus from the old nest to start a new crop.

The ants and the fungus have formed a lasting bond of mutual dependency. Without their fungus crop the ants would die. Without ants to cultivate the fungus, it would die. The lives of the partners are so intertwined that when a biological change happens in one of them, it is accompanied by a change in the other. This is called coevolution. Ants and their fungus crop have coevolved for millions of years. How do scientists know this? They

Cameron Currie and colleagues search tropical forests for ant farms, which they bring back to their laboratories to study.
Photo courtesy Anna Himler

recently discovered the long history of this relationship by tracing the ancestry of the ants and their fungus crop through their DNA. The fungal crops found on ant farms today descended from a single fungus that lived about 50 million years ago. As the ants branched into new species, the fungus also branched into new species.

Scientists once believed that the ants were successful for so long because they

The ant, *Acromyrmex,* tending its fungus crop.
UNSM Angie Fox illustration

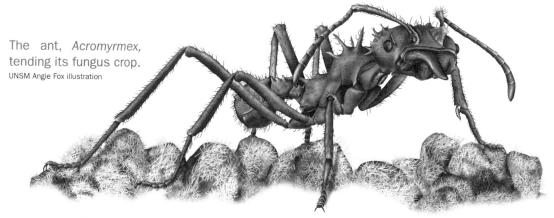

kept their fungus crop free of disease. Ants are meticulous farmers. They lick the surfaces of their nest and each other with their tongues to keep everything spotless. They reduce the number of pests entering the farm by chewing the harvested leaves into mush. They weed the garden carefully, removing weeds to special dump chambers or to garbage heaps outside the nest. The research suggested that the ants were so good at keeping out pests that no disease could take hold.

Among the ants that Cameron Currie studies are two types of leaf-cutter ants, *Atta* and *Acromyrmex* (ack-crow-MUR-mex). As a young graduate student, Cameron had a hunch that these and other farming ants did have a pest problem; it just hadn't been discovered. He based his hunch on examples of human crop disasters. Anyone who farms knows that farmers are constantly battling a barrage of pests. And the battle gets tougher when farmers grow the same crop year in and year out. The crops usually do well for a while, but they are sitting ducks for diseases. Was life really any different on ant farms?

Cameron set off for the tropics to collect ant nests to study. He searched under rocks and in ravines for nests small enough to pack up and take back to his lab. In his first summer of research he collected between 50 and 60 walnut-sized nests from various fungus-growing ant species. In the lab, after the ants had time to settle down from their trip, he looked for evidence of infection by crop pests.

Cameron's hunch proved right. A crop pest called *Escovopsis* (es-co-VOP-sis) kept showing up in the ant farms. This pest is found only on the fungus crop of farming ants. This pest is found nowhere else in nature, yet is extremely common on the ant farms. And each ant species that Cameron looked at had its own specialized species of *Escovopsis*. All of this suggested to Cameron that the crops pest had coevolved with the ant and the fungus crop.

Cameron also discovered that the crop pest could be deadly. If he removed the ants from a farm infected with the pest, the crop died in just a few days. But if the ants were allowed to stay, the pest only slowed the growth of the crop but didn't wipe it out. What secret did the ants have that kept the deadly crop pest under control?

Cameron began to wonder about a white coating he noticed on many of the ants. The ants working deep in the nests were often covered with it. When he examined the substance under a powerful microscope, he discovered it was actually alive. The ants were wearing a coating of tiny bacteria. Were they harmless hangers-on—like fleas on an elephant—or was this news? Did the bacteria help the ants protect their fungus crop from the pest?

Through experiments, Cameron discovered the key to the smooth-running ant farms. The bacteria living on the ants produce antibiotics that kill the crop pest and keep the ant farms running well. The bacteria are known as actino-mycetes (ak-TIN-oh-my-see-tees). They also produce many of the antibiotics that humans use to fight disease. But ants used them first!

Explore how Cameron set about discovering some of the fascinating interactions on a tiny farm cooperative. In this activity you'll investigate life on a walnut-sized farm, simulate petri dish experiments in which bacteria are introduced to killer pests, and decide whether the results of the experiments provide evidence for coevolution.

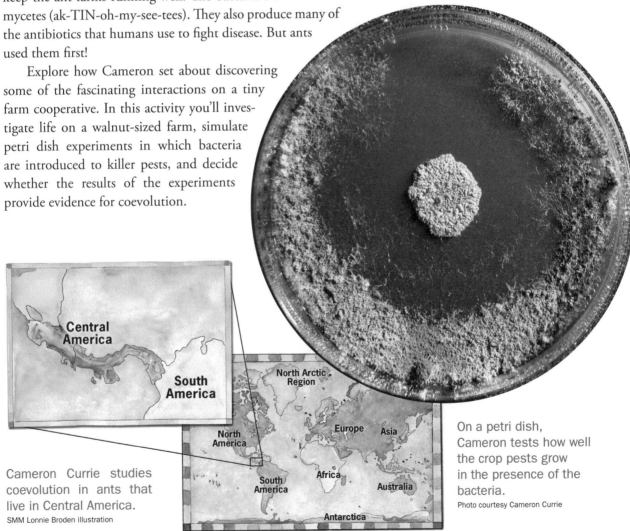

On a petri dish, Cameron tests how well the crop pests grow in the presence of the bacteria.

Photo courtesy Cameron Currie

Cameron Currie studies coevolution in ants that live in Central America.

SMM Lonnie Broden illustration

PART ONE
Three buddies & a menace

You're invited to investigate the underground world of the leaf-cutter ants and their partners, the fungus crop and bacteria . . . and their enemy, the crop pest *Escovopsis*. First, you will make a set of cards telling about each player among the leaf-cutter ants' partners. Then create a poster showing how they work together.

Work with a partner

Each team of two will need:

- Ant Fact Cards 1 and 2
- tape or glue stick
- large sheet of paper (about 11x17)
- scissors

1 Meet the Players

a Cut out the four cards along the broken lines. Fold each card on the solid line so you can read the writing. Paste or tape the cards closed.

b Now use the cards to create a poster showing the leaf-cutter partners. Read the information on each card to help you decide where to place the card. Tape the top of the cards to a sheet of paper in a pattern you think best shows the relationships connecting the players (who depends on who). Make sure you can flip the card over to read the other side.

2 Chart Their Game

Draw arrows on the poster between the players showing the relationships. Title the poster.

Farming Ant
Acromyrmex
(ack crow MUR mex)

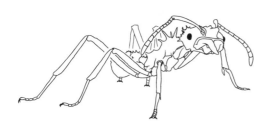

Size: 1-2 centimeters
length of a zipper pull

Fungus Crop
Lepiotaceae
(Lep ee oh TAY seeah)

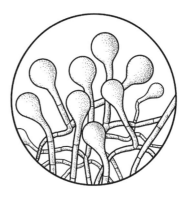

Size: 30-50 microns
Two strands equals a human hair's width.

Role: Farming Ant
Name: *Acromyrmex*

Habitat: *Acromyrmex* ants live in nests about 50 centimeters deep. About 500 to 100,000 worker ants live in a single nest.

Habits: These ants can't eat the leaves in tropical forests because they contain poisons. They have evolved ways to use these toxic leaves to grow their own food. Big ants harvest the leaves, and smaller ants chew them into a paste. The smallest ants spread the paste in the nest chambers so the fungus will grow. Ants weed and prune their crop to keep it healthy. The fungus crop is the ant's only food.

Role: Fungus crop
Name: *Lepiotaceae*

Habitat: This fungus grows in *Acromyrmex* colonies. It lives off leaf mulch supplied by the ants.

Habits: This fungus looks like tiny strands with swollen tips. The tips are highly nutritious and the ants digest them easily. This fungus cannot reproduce on its own. It depends on the ants to be cultivated and supplied with leaves.

UNSM Angie Fox ant illustration, Cara Gibson Escovopsis.

3

Crop Pest
Escovopsis
(es co VOP sis)

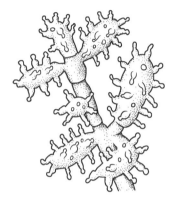

Size: 100 microns
width of a human hair

Bacteria
Actinomycetes
(ac TIN oh my see tees)

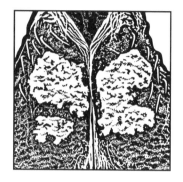

Size: 1 micron
2,000 fit across the head of a pin.

Role: Crop Pest
Name: *Escovopsis*

Habitat: This crop pest is a hair-like fungus that is found in nests of *Acromyrmex* ants. It is a parasite that lives off the ants' fungus crop. When it overwhelms the fungus crop, the ants starve and die.

Habits: The crop pest looks like a white cloud when it blooms. Ant nests often start off free of the crop pest, but within two years most of the ant nests become infected.

Role: Bacteria
Name: *Actinomycetes*

Habitat: These bacteria live on the surface of *Acromyrmex* ants. They appear on the underside of the ant like a white dust. Sometimes this dust completely covers an ant's body.

Habits: The bacteria produce an antibiotic that helps the ants protect their farm from the crop pest. Glands inside the ant's body appear to nourish these bacteria.

3 Consider This

a   Make an award in the Leaf-Cutter Game of Life. Who is the most valuable player? Why?

b   You're the ref: Which player can you remove without stopping the game?

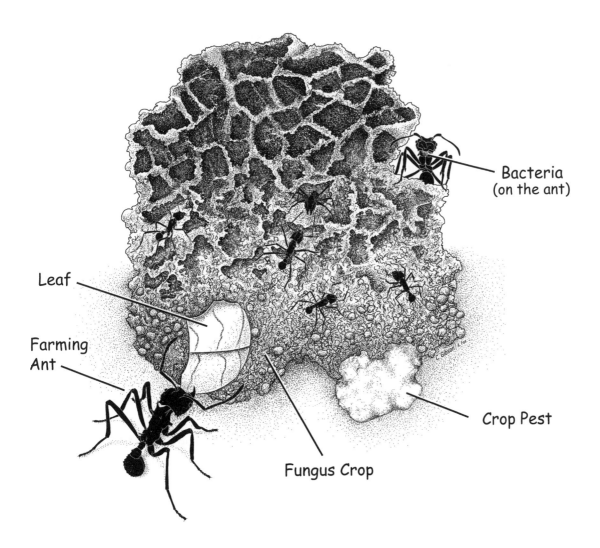

Leaf

Farming Ant

Fungus Crop

Bacteria (on the ant)

Crop Pest

PART TWO
Killers vs. the bacteria

After investigating hundreds of ant nests, Cameron discovered the ants have a serious pest called *Escovopsis* that kills their fungus crops. When the ants were around, the crop pest seemed not to cause much damage to the fungus crop. But when the ants were taken from the nest, the pest could kill a fungus crop in just a few days.

The ants appeared to protect their crop from the crop pest, but how did they do it? Cameron discovered that there is a patch on the ant's body that supports a thick concentration of bacteria. Close inspection with a microscope showed that under this patch there is a gland that secretes substances that nourish the bacteria (called actinomycetes). So the ants appeared to be feeding the bacteria. But why? What do these bacteria do for the ants?

Cameron guessed that the bacteria help the ants protect their crop from the pest. To test his idea he began growing samples of the crop pest and testing the effects of the bacteria on the pest's survival. Check out Cameron's experiment, measure his results, and then come to your own conclusions about the effects of the bacteria.

Work with a partner

Each team will need:

- Crop Pest vs. Bacteria Experiment
- Crop Pest vs. Bacteria Data Chart
- cm ruler
- two colored pencils (red plus one other)

1 Set-Up: Control and Test Experiment

In the lab, Cameron grows the crop pest in petri dishes. Petri dishes are sterile round containers with tight-fitting lids. The lids prevent other micro-life from getting in and contaminating his experiments. They also keep the crop pest from getting out and contaminating his lab. To begin, Cameron pours a layer of food material in each petri dish so the crop pest has something to feed on.

To find out whether the bacteria prevent the crop pest from growing, Cameron sets up a control and test condition.

a   A control is the part of the experiment that shows what happens when you don't change or add anything extra. For the control, what should be in the petri dish?

b In this test, the crop pest grows on a petri dish that is like the control, but with one difference: A spot of bacteria is added in the middle of the dish. What would you expect to find if the bacteria stop the pest?

2 Running the Experiment

Cameron introduces a small amount of the crop pest at the edge of the dish. This is the control. Cameron places the dishes in a warm place and records the growth on day 1, 5, 10, and 20.

a Look at the Crop Pest vs. Bacteria Experiment sheet. On the section labeled "control," color the crop pest (the growing spot at the edge of each dish on days 1–20) one color (but **not** red). Notice how the crop pest changes from day 1 to day 20.

Crop Pest vs. Bacteria Experiment

CONTROL: Crop Pest *Escovopsis*

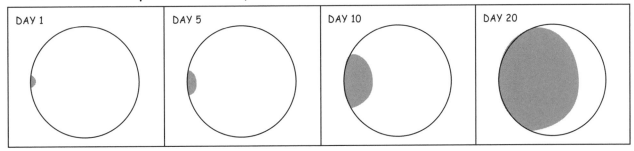

TEST: Crop Pest *Escovopsis* + Bacteria *actinomycetes*

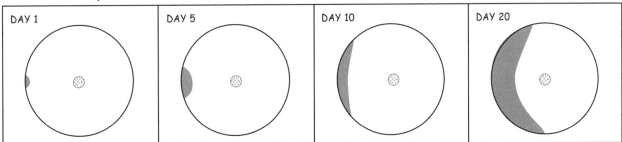

b On the same experiment sheet look at the section labeled "Test." Here Cameron introduces the crop pest at the edge of dish and a spot of the bacteria in the center. Again, Cameron places the dishes in a warm place and records the growth on day 1, 5, 10, and 20.

c On the experiment sheet, color the bacteria red. Color the crop pest the same as in the control. Notice how the pest changes from day 1 to day 20.

3 Measure and Chart the Results

a Measure the growth of the crop pest in each dish in millimeters. Start from the round edge of the petri dish and measure toward the center of the dish.

b Record the measurements for each dish on the Crop Pest vs. Bacteria Data Chart.

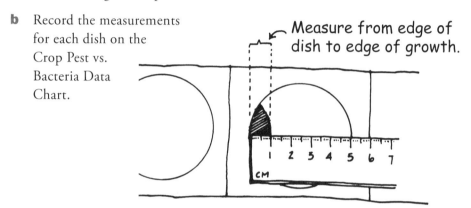

Measure from edge of dish to edge of growth.

Crop Pest vs. Bacteria Data Chart
Measuring the effects of bacteria on the growth of a crop pest

	DAY 1	DAY 5	DAY 10	DAY 20
CONTROL Crop Pest alone (*Escovopsis*)				
TEST Crop Pest + Bacteria (*Escovopsis* + actinomycetes)				

4 Consider This

Review your data. What can you conclude about the effect of the actinomycetes bacteria on the crop pest *Escovopsis?*

PART THREE
Evidence for coevolution

In coevolution, changes in one species are accompanied by changes in another. Many scientists have established that the ants and their fungus crop are a good example of coevolution, because the two are completely dependent on each other for their survival.

Cameron Currie began to wonder if the bacteria and the crop pest might also have coevolved with the ant and the fungus crop. This would mean a four-partner coevolution—not just two partners. But how could this coevolution be determined?

Cameron decided to test whether the bacteria affected the growth of only the pest *Escovopsis*, or whether it affected other fungus species as well. If the bacteria developed a substance that would kill only one kind of crop pest, then that would support the idea that the two organisms had coevolved.

Work with a partner

Each team will need:
- Crop Pest vs. Bacteria vs. New Pest Experiment
- Crop Pest vs. Bacteria vs. New Pest Data Chart
- cm ruler
- three different colored pencils (red plus two other colors)

1 Controls and Tests

As in the first experiment, the microorganisms Cameron tests grow in petri dishes. To begin, Cameron pours a layer of food material in the 1, 2, and 3 petri dishes so the microorganisms in each dish have something to feed on.

a To test whether the bacteria can kill other kinds of crop pests, Cameron repeats the previous experiment with some additions. The new test condition is added to see whether the new kind of crop pest will be stopped by the bacteria.

b For this test to be convincing, three controls are needed. Look at the Crop Pest vs. Bacteria vs. New Pest Experiment sheet to see the controls. Name them here:

Control 1:

Control 2:

Control 3:

Crop Pest vs. Bacteria vs. New Pest Experiment

CONTROL ONE: Crop Pest *Escovopsis*

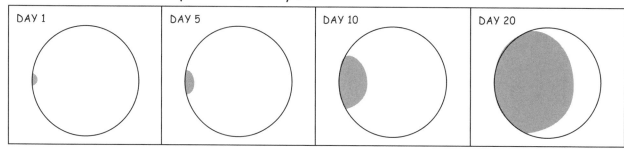

CONTROL TWO: Crop Pest *Escovopsis* + Bacteria actinomycetes

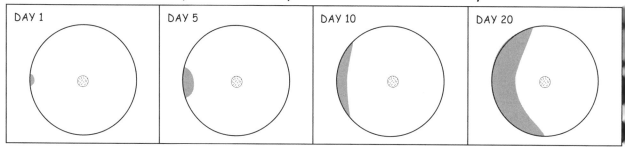

CONTROL THREE: New Pest *Trichoderma*

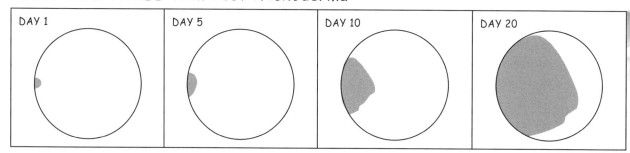

TEST: New Pest *Trichoderma* + Bacteria actinomycetes

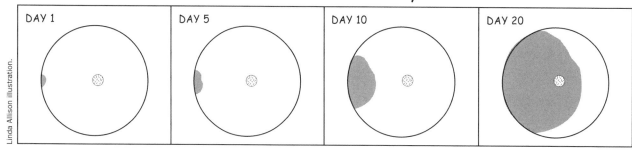

Linda Allison illustration.

Ants & Co.: Tiny Farms

c   Why are these conditions considered controls for the test condition?

2 Running the Data

Repeat the procedures as in the previous experiment. On the Experiment Sheet dishes where bacteria are added, color the bacteria red. Color the crop pest *Escovopsis* a different color. Color the new pest *Trichoderma* a third color.

3 Measure and Chart the Growth

Measure the growth of the crop pest in each dish in millimeters. Start from the round edge of the petri dish and measure toward the center of the dish. Record the measurements for each dish on the Crop Pest vs. Bacteria vs. New Pest Data Chart.

Crop Pest vs. Bacteria vs. New Pest Data Chart

Measuring the effects of bacteria on two different kinds of crop pests

	DAY 1	DAY 5	DAY 10	DAY 20
CONTROL ONE Crop Pest alone (*Escovopsis*)				
CONTROL TWO Crop Pest + Bacteria (*Escovopsis* + actinomycetes)				
CONTROL THREE New Pest alone (*Trichoderma*)				
TEST New Pest +Bacteria (*Trichoderma* + actinomycetes)				

4 Consider This

Why did Cameron test the bacteria with the new pest?

Hint: A defender that kills one and only one kind of pest may have coevolved with it.

Do you think this experiment suggests that the bacteria and the crop pest may be coevolved partners? Why or why not?

PART FOUR
Be a science reporter

Write a short news story about ants and their partners. Tell your readers about how leaf-cutter ants, their fungus crop, the crop pest, and the bacteria have lived together in an association for millions of years. Based on what you have learned, explain how you think this association came about. What evidence supports your explanation?

P.S. Don't forget the headline.

HAWAIIAN FLIES:
Song & Dance Success

We may think only humans are sexy, but tell that to a Hawaiian fly when he's waggling his bottom, whirring his wings, and trying his hardest to attract a mate. Like most animals, Hawaiian flies spend a lot of their energy engaged in the mating game. Their sex lives are hidden from us only because of their tiny size.

Fly sex is no secret to Ken Kaneshiro, a biologist at the University of Hawaii. With the aid of a microphone, a video camera, and a device that records sound vibrations, Ken studies the mating behavior of Hawaiian *Drosophila* (dro-SOFF-ih-lah), a group of flies found only on the Hawaiian Islands. These flies are related to *Drosophila* flies worldwide, including the flies that hover over ripe bananas in your fruit bowl. But Hawaiian *Drosophila* aren't ordinary flies. They're famous for having some of the most diverse and bizarre courtship behaviors in the world.

There's *Drosophila silvestris* (sil-VEST-ris), a flashy fly with spotted wings. To attract a female

This Hawaiian *Drosophila (Drosophila differens)* is perched on a decaying Clermontia shrub. Female *D. differens* flies lay their eggs on Clermontia branches. When the eggs hatch, the larvae feed on the shrub as they develop into adults.
Photo courtesy Kevin T. Kaneshiro

he performs a kind of solo tango, sliding back and forth and circling her in a series of intricate steps while vibrating muscles in his abdomen to serenade her with purring sounds. If he gets tired, she chases after him and hits him with her legs until he starts dancing and singing again. Then there's *Drosophila heteroneura* (heh-ter-OH-noora), a fly with a big head shaped like a double hammer. To impress a mate he uses his head to fight other males who enter the lek (an arena where males go to strut their stuff). And then there's *Drosophila glabriapex* (gla-BREAH-pex), a fly who romances his mate by rubbing his hairy legs on her abdomen.

Ken Kaneshiro and other fly researchers have identified at least 800 species of *Drosophila* in Hawaii, each of them strikingly different. No other place in the world has so many diverse fly species in so small a space. And what is more astonishing, it probably started with just one pregnant fly that blew ashore several million years ago. Biologists call the explosive evolution of one or two species into hundreds "adaptive radiation." The Hawaiian *Drosophila* are one of the world's foremost examples. "What is it about the Hawaiian Islands," Ken wonders, "that has contributed to so many species?"

To understand what is special about Hawaii, start by imagining the immense Pacific Ocean. Then imagine a small chain of volcanic islands in the middle of it, more than two thousand miles from the nearest continent. For a tiny fly to make it from North America or Asia all the way to one of these islands (long before ship or plane travel) is remarkable in itself. Ken and other scientists speculate that she was blown there by a storm several million years ago.

Hawaii is remote, but it is also a tropical paradise rich in diverse habitats. A fly that managed to survive the journey was a lucky fly. Her new surroundings offered opportunities she didn't have back on the continent, where the environment was already crowded with competing insects and voracious predators. Hawaii is so isolated that only some kinds of organisms can make it there. The first fly and her small brood were safe from most competitors and predators. Some of her offspring thrived in the new environment and mated, and the tiny population built up.

Because of their size, *Drosophila* flies don't usually travel far. They tend to live generation after generation in the same neighborhood. But a few flies from the first tiny population, or their descendants, traveled on hurricane winds to other islands, or to different habitats on the same island.

To a fly, even a tiny Hawaiian island has many small, isolated habitats, ranging in extremes from lush mountain forests to barren lava fields. Not all the flies that landed in these strange new habitats were able to survive and reproduce. Some of them had features or behaviors better suited to finding food or laying eggs in their new surroundings. These flies were more likely to survive and produce more offspring. The more suitable characteristics passed to the next generation, and gradually throughout the population. This process is called evolution through natural selection. Hawaiian *Drosophila* populations evolved unique adaptations to the different places they landed in. Today, for example, there are species of flies that hunt prey such as spider eggs, others that sip flower nectar, and still others that eat only rotten food. And there are different species that lay their eggs in rotting leaves, bark, fruit, and fungi.

But how did different fly populations become new species? Over thousands of generations, as the flies in isolated new habitats became more and more different from populations in older habitats, they eventually became unable to recognize or breed with the ancestral populations. They had evolved into a new species.

This is where fly sex comes into the picture. Having the ability to survive is only half the story. Mating is the other half. And attracting a mate is a big part of that story. Scientists call this principle of evolution sexual selection. Many organisms have characteristics and behaviors that serve no other purpose than to attract a mate. Male peacocks, for example, have large colorful tails that they display to catch the eye of females. The larger and more colorful the tail, the more likely the peacock is to mate and produce offspring. The offspring with the largest, most beautiful tails are also most likely to mate and reproduce, ensuring that future generations of peacocks evolve even larger and more colorful tails—at least up to a point. If a tail is too large and heavy it could be a hazard rather than an asset, making it difficult for the peacock to escape from predators and reducing his chances of surviving and reproducing.

Ken Kaneshiro and other researchers have observed the principle of sexual selection at work in Hawaiian *Drosophila*. Body features such as bold wing markings and the amount of hair on the legs, as well as behaviors such as intricate songs and dances, often lead to mating success. These sexual characteristics and behaviors are tightly controlled by genes, which means that "sexiness" is inherited. A female fly that is attracted to a hairy-legged male and mates with him will tend to have hairy-legged, and therefore sexy, sons. Like their hairy dad, the sexy sons will attract females that share

Ken Kaneshiro from the University of Hawaii studies the evolution of Hawaiian *Drosophila*. Ken smears rotten bananas on trees and shrubs to attract his research subjects.

Photo courtesy Robert Chin, University of Hawaii

Virus and the Whale: Exploring Evolution in Creatures Small and Large

Two male Hawaiian flies *(Drosophila heteroneura)* in the head-to-head posture of their territorial defense display.
UNSM Angie Fox illustration

their mother's tastes.

Ken studies sexual selection in Hawaiian *Drosophila* by tracking different species in the wild, trapping them with bait, and raising them in the lab. He takes males and females, puts them together in a mating chamber, and observes and records their behavior. He keeps track of which males and females will mate and pass on their genes.

Ken has discovered that in any population of flies, some males are great performers and some are lousy. He also observed two types of females—picky and passionate. The picky females only mated with a male if they recognized and liked his performance. The passionate ones would take any fly that came along, whether he performed well or not. Ken was one of the first fly researchers to study female mating preferences and the difference their choosing makes to future generations of flies.

From his observations, Ken has developed a theory. He believes that female choosiness has driven the rapid evolution of fly species in Hawaii—and some very bizarre body shapes and courting behaviors. The picky females play an important role in sexual selection when fly populations are large and there are lots of males to choose from. But passionate females play an even more important role, especially in small populations when mate choice is slim. In those conditions picky females may not mate at all, and passionate females are likely to choose "whatever comes along," including flies with bizarre new features and behaviors.

Although you may think flies are an unusual case, the evolution of elaborate sexual characteristics is really very common. Investigate the fascinating realm of sexual selection for yourself, and consider the effects of what happens when females call the shots. In this activity you'll meet some interesting flies, learn how to sing their songs, and create a model to test the theory that passionate rather than picky females have the advantage in reproducing.

PART ONE
Meet the flies

You probably have already met a *Drosophila*. These little flies are the ones you find lazily buzzing around a fruit bowl especially when it is warm. You have probably never given them a second thought… other than "what are those weird little bugs doing on my bananas?" There are people, however, who spend most of their time thinking only about flies. *Drosophila* are perhaps the most studied of all animals. People know more about fly genetics and development than they know about humans. But why would anyone care about flies? Lots of reasons. Shuffle through the fly fact cards and find out.

Work with a partner

Each team will need:

- Fly Fact Cards 1–3 (cut, fold, and tape the 12 fly cards)
- paper
- tape or glue
- scissors

1 Fly Fact Cards: What Sort?

a Cut the fly cards apart along the dotted lines so the text and the image are connected. Fold the cards along the center line. Tape or glue them closed so they make a two-sided card.

b Sort the fly card pictures into groups or families in a way that makes sense to you. What feature or features did your group use to sort?

c Share your sorting with the other groups. Did everyone sort the same?

2 Consider This

Read the fly facts on the other side of the cards (the facts don't necessarily match to the fly picture). Choose your favorite fly fact. Each person writes his or her fact as a question on a slip of paper. Wad the slip up and throw it into the middle of the table. Draw a question (one you didn't write). See if you can remember the answer.

Fly Fact Cards 1

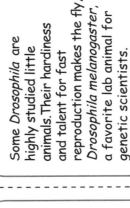
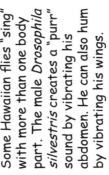

Drosophila fasciculisetae

Some *Drosophila* are highly studied little animals. Their hardiness and talent for fast reproduction makes the fly, *Drosophila melanogaster*, a favorite lab animal for genetic scientists.

Many Hawaiian *Drosophila* are still mysterious creatures, yet to be studied and named.

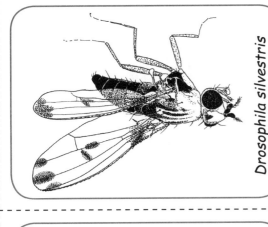

Drosophila silvestris

Some Hawaiian flies "sing" with more than one body part. The male *Drosophila silvestris* creates a "purr" sound by vibrating his abdomen. He can also hum by vibrating his wings.

Some females prefer a male who sings while holding his head under her wings.

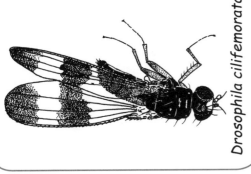

Drosophila cilifemorata

Tail wagging, big antennae, extra hairy legs, songs and dances are features that female Hawaiian *Drosophila* find sexy. Females of each species are attracted to only some features and not others.

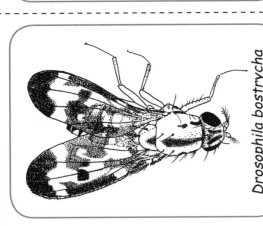
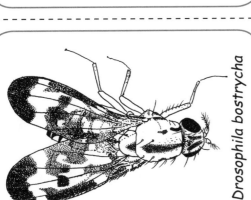

Drosophila bostrycha

Some Hawaiian *Drosophila* are called the "Birds of Paradise" of the insect world because of their spectacular courting displays.

Francisca C. do Val Illustration.

Virus and the Whale: Exploring Evolution in Creatures Small and Large

Fly Fact Cards 2

Drosophila heteroneura

The youngest Hawaiian island, the Big Island, is only half a million years old. The 26 species of picture winged *Drosophila* on the Big Island evolved within that time span.

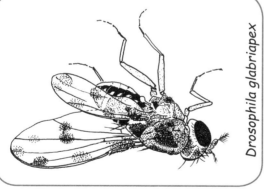

Drosophila grimshawi

The Hawaiian islands are the world's most isolated landmass. Two thousand miles of ocean on all sides surround them. It is amazing for any insect to have arrived here.

Drosophila glabriapex

Drosophila are found all over the world. Hawaii is home to a quarter of the world's species in an area no bigger than the state of Connecticut.

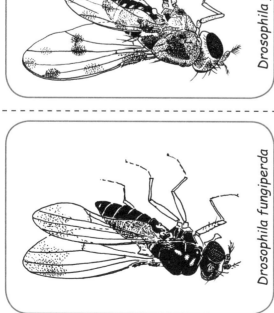

Drosophila fungiperda

There are over 800 species of *Drosophila* in Hawaii. Ninety percent of Hawaii's species are endemic (found nowhere else in the world).

Fly Fact Cards 3

✂ Cut on dashed lines, fold on solid line.

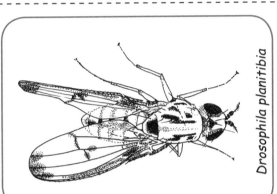

Drosophila crucigera

A male of one species of Hawaiian *Drosophila* sings and dances with a series of virtuoso steps, circling, gliding back and forth, whirring his wings, waggling his body.

A male may repeat a dance for an hour or more to please a female. If the male finally gives up, the female may rush after him and rear back, hitting him with her front legs. That makes him come back and start all over again.

Drosophila planitibia

Compared to *Drosophila* on the U.S. mainland, those on Hawaii have low reproductive rates. Some Hawaiian *Drosophila* species lay only a few eggs in their lifetime, while others may lay up to 200 eggs.

Drosophila pilimana

All Hawaiian *Drosophila* may be descended from a single pregnant female. Scientists think she may have blown ashore about eight million years ago.

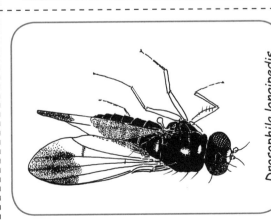

Drosophila longipedis

Many *Drosophila* have orange-brown bodies and measure from 1/16 to 1/8 inches in length. These flies have clear, unmarked wings.

Over 100 species of Hawaiian *Drosophila* are larger, up to 3/4 inches in length with bold black wing markings.

Francisca C. do Val illustration.

PART TWO
Rate-a-mate

Simply surviving is only part of being successful. Getting your genes into the future population is also a necessary part of survival. Finding a receptive mate of your own species is key to reproductive success. In Hawaiian *Drosophila* signature songs and dances help males and females of the same species find each other. Female flies choose mates only when they recognize their species' special song and if they like the male's performance. In the fly world, it's ladies' choice.

Dr. Ken Kaneshiro and his team at the University of Hawaii have made a study of fly songs. This is not easy since the fly tunes are not really audible to human ears. A special microphone is placed close to a pair of courting flies to record the songs. The recordings are graphed as pulse patterns on an oscilloscope.

What have the researchers found? *Drosophila* males sing by fanning or vibrating their wings. The vibrations are very fast (measured in thousands of beats per second). If you were to listen you wouldn't call it music because there is no tune. It is more like pulses or beats. However, each pulse pattern is unique to a species of fly. How well a male scores in the mating game depends on what kind of pulses he puts out. Now it's your turn to investigate some fly songs.

Divide the players into two groups. Each group should have the same number of players. One group will be female flies and the other will be male flies.

Each player will need:
- Fly Song Simulator
- Songs of Hawaiian *Drosophila* card, either a male or a female, depending on your group.
- chair
- scissors

1 Make a Fly Song Simulator

a   Cut the song simulator out along the dotted lines.

b   Fold it in half on the dotted line. Cut the diamond shape out of the center.

c   Fold the flaps back to make handles.

Snip

fold

humm

z z z z z

Linda Allison illustration

2 Make Your Simulator Sing

a Hold the flaps between two fingers about 1/8 inch apart and press them to your lips. Blow through your mouth so that air blows out through the hole. The paper will vibrate. What does it sound like? (If this doesn't work, try blowing and humming a tone at the same time.)

b Use your song simulator to play some of the songs from the Songs of Hawaiian *Drosophila* Cards.

- Use one tone.

- Vary the vibration or buzz by changing the loudness (tall lines) or softness (short lines). Try blowing harder and softer or pinching the flaps closer or farther. Experiment!

- Vary the length of time that you hold a tone.

FLY SONG SIMULATORS

✂ Cut on dashed lines, fold on solid lines.

✂ Cut on dashed lines, fold on solid lines.

✂ Cut on dashed lines, fold on solid lines.

✂ Cut on dashed lines, fold on solid lines.

Linda Allison

SONGS of HAWAIIAN *DROSOPHILA* - Female Cards

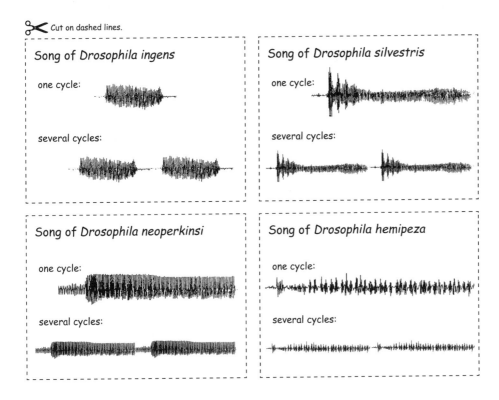

✂ Cut on dashed lines.

Song of *Drosophila ingens*

one cycle:

several cycles:

Song of *Drosophila silvestris*

one cycle:

several cycles:

Song of *Drosophila neoperkinsi*

one cycle:

several cycles:

Song of *Drosophila hemipeza*

one cycle:

several cycles:

SONGS of HAWAIIAN *DROSOPHILA* - Male Cards

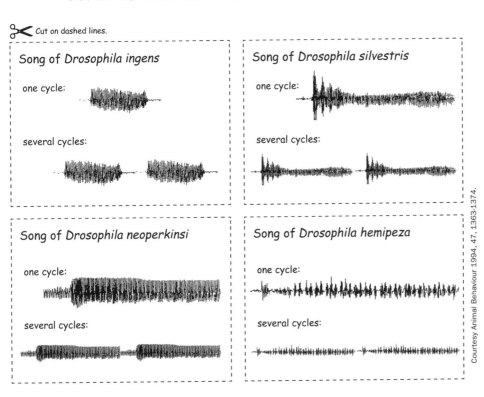

✂ Cut on dashed lines.

Song of *Drosophila ingens*

one cycle:

several cycles:

Song of *Drosophila silvestris*

one cycle:

several cycles:

Song of *Drosophila neoperkinsi*

one cycle:

several cycles:

Song of *Drosophila hemipeza*

one cycle:

several cycles:

Courtesy Animal Behaviour 1994, 47, 1363-1374.

3 Play Rate-a-Mate

a Cut out the song cards. Create a deck of four male cards and four corresponding female cards. Pass out the cards to the males and females. Each card tells you what kind of fly you are and your special song. Keep your cards secret!

b Male flies should take a few minutes to practice performing the song on their card using their simulators. Meanwhile female flies should be setting up pairs of chairs around the room; each pair should be back-to-back facing opposite directions.

c When you are ready to begin, each male fly picks one of the empty chairs in the pairs. This will be their "lek" or territory. Male flies defend their territories or leks from other males.

d Female flies migrate around the room and pick an empty chair next to a male fly. When she settles down, the male fly performs the song on his card. Be sure you are back-to-back and you can't see each other's cards, but you are close enough to hear the song.

e Male flies perform the song and the female fly listens. When she has heard enough, it is up to her to choose if she likes the performance, and if the song matches her species (the song on her card).

f Play enough rounds so all the players have a chance to pair up. If there is time you may want to switch roles and play another round. Make new simulators if someone new is going to use them.

4 Consider This

a Did you find a mate of your species?

If not, do you think failing to find a mate was because the females were too choosy? Or was it because the male songs were performed badly?

b Would you say that easy-to-please females or choosy females have a better chance of finding a mate?

Would you think that males that are poor performers or good performers have a better chance of finding a mate? Why?

PART THREE
Song-and-dance success

Do you think choosy female flies or non-choosy females are going to have better reproductive success over several generations? Experiment with a model of sexual selection and test your prediction.

Work with a partner

Each team will need:
- Fly Paper sheet
- Tracking Fly Population Chart
- scissors

1 Create a Population

Cut out all the squares on the fly paper. Clear a tabletop space. Sort the different flies into piles. Start by counting out a population of the following: three Picky Females, three Passionate Females, three Cool Males, three Klutz Males.

Picky Females These choosy females mate only with cool males . . . good singers and dancers.	**Cool Males** These males are good singers and dancers.
Passionate Females These non-choosy females mate with any male.	**Klutz Males** These males are poor singers and dancers.

2 Make a Prediction

Picky or Passionate? Will Picky (choosy) or Passionate (non-choosy) females have better chances of mating?

FLY PAPER

COOL MALES: good singers and dancers

PICKY FEMALES: only mate with cool males

KLUTZ MALES: poor singers and dancers

PASSIONATE FEMALES: mate with any male

Hawaiian Flies: Song & Dance Success

Linda Allison illustration

3 Ready, Set, Experiment!

a **Scramble:** Line up the six female flies in a random order (Female flies are on a white background). Line up the six male flies at random, opposite the females (Male flies are on a toned background).

b **Mate:** Bring the two lines together to make pairs of males and females. In the real world, now is when the male flies begin their courtship songs and dances. Picky females will not mate with a male that can't sing or dance. Any pair of picky female and klutz male is doomed. Remove them both.

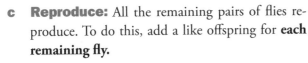

c **Reproduce:** All the remaining pairs of flies reproduce. To do this, add a like offspring for **each remaining fly.**

d **Count:** Tally the fly population by type. Record your findings on the Tracking Fly Population Chart.

e Repeat the mating cycle using only the remaining flies. Do this for generation 2, 3, and 4. Go through **all** the steps for each generation: Scramble, Mate, Reproduce, Count. After each mating, count the flies by type and record the results for each generation on the Tracking Fly Population Chart.

Tracking Fly Population Chart

FLY TYPE	Generation 1	Generation 2	Generation 3	Generation 4
PICKY FEMALE	3			
PASSIONATE FEMALE	3			
COOL MALE	3			
KLUTZ MALE	3			
TOTAL POPULATION (add all Males and Females)	12			

NOTE: Pairs of picky females & klutz males do not produce offspring.

Linda Allison illustration

4 Check It Out

a Which male flies were the most successful reproducers in your experiment?

b Which female flies were the most successful reproducers?

b Check your results with another team. Did their results agree or disagree with yours?

5 Consider This

Based on your results, which males and which females are more likely to pass their genes into the future? How would this affect a fly population over time?

PART FOUR
Be a science reporter

Write a short news story about Hawaiian flies. Tell your readers about how eight million years ago, there were no *Drosophila* flies on Hawaii. Now there are more than 800 species found only on the islands. Based on what you have learned, explain how you think so many new species came to be on the islands.

P.S. Don't forget the headline.

GALÁPAGOS FINCHES:
Famous Beaks

Why would anyone travel halfway around the world to a hot, rocky island to measure bird beaks? Year after year, for more than 30 years, biologists Rosemary and Peter Grant have returned to the Galápagos Islands to do just that. These aren't just any islands, and the birds aren't just any birds. The Galápagos Islands are a cluster of active volcanic islands in the Pacific Ocean. The island chain is extremely remote. It lies hundreds of miles from the nearest landmass, South America. The plants and animals that colonized the islands evolved in unique ways. The islands, and especially a group of dull little birds known as Darwin's finches, helped inspire Charles Darwin's ideas about evolution. The finches are named in his honor.

The Galápagos Islands are home to 13 species of finches. The birds vary in size and shape, and they are about as drab looking as a sparrow—until you consider their beaks. Each species is distinguished by a different beak. The 13 beaks are often compared to a toolbox containing 13 kinds of pliers, each one suited for a different task. The cactus finch has a long thin beak that works well for crushing cactus seeds. The woodpecker finch uses its beak to dig out insects from dead wood. The sharp-beaked ground finch feeds on nectar from flowers and catches insects, but sometimes it also pecks through the skin of seabirds called boobies and feeds on their blood. While the finches on the Galápagos feed on many kinds of foods, each bird has a beak suited to certain eating habits.

When Darwin first saw the finches on his visit to the Galápagos in 1835, he didn't know how important they would be to him. Darwin was 26 years old and employed as a naturalist on a five-year voyage around the globe. One of his jobs was

Rosemary and Peter Grant study the evolution of finches on the Galápagos Islands. They are professors at Princeton University.

Photos courtesy Rosemary and Peter Grant.

to collect samples of unknown plants and animals. Darwin captured some of the Galápagos finches for his collection, but at the time, he was more interested in plants and rocks and in the islands' stranger inhabitants. In his diary he mentioned the "hideous" lizards that gathered seaweed in the ocean and giant tortoises that were big enough to ride on.

Years later, however, it was the collection of little finches that puzzled and inspired him most. Darwin wondered why there were so many species of finches on the Galápagos Islands, and why they were as different and as similar as they were. Finches don't migrate, so the birds he collected must have evolved on the islands. He speculated that the first birds blew to the Galápagos from the coast of South America, evolving in a dozen directions on different islands.

The finches' beaks gave Darwin a clue about how a species could evolve. The size and shape of a bird's beak determine the kinds of food the bird can eat and the kinds it can't. A slight difference might give one bird an advantage over another in surviving and reproducing, and the advantaged offspring, in turn, would be more likely than others to survive and reproduce. Darwin called this process natural selection. Others described it as "survival of the fittest." Darwin thought that natural selection worked too slowly to be seen in one's lifetime.

More than a hundred years after Darwin's visit, Rosemary and Peter Grant traveled to the Galápagos Islands to take a closer look at the finches. The Grants wondered: If they took careful measurements of the finches and the foods they ate, would they be able to see the changes that Darwin imagined?

The Grants and their students set up a research camp on Daphne Major, an island in the center of the Galápagos. Daphne Major is a biologist's dream because of its small size, isolation, and harsh weather. The island is like a giant petri dish, a science experiment set up by nature.

Daphne Major has a resident population of finches commonly known as medium ground finches. Their scientific name is *Geospiza fortis* (geo-SPEEZA-fortiss). They are distinguished from other finch species by their medium-sized bodies and their beaks. When the Grants started their research, they took a head count of the finches on Daphne Major. To do this, they put up mist nests to catch the birds, picking them out one by one with their hands and placing numbered bands on their legs. Sometimes the finches watched, perching on the researchers' wrists and arms as they worked.

After banding the birds, the researchers took careful measurements. They weighed each bird with a spring balance and then measured it from head to hallux (big toe) with a set of calipers. They kept a record of the bird's wing length, leg length, toe length, and, most important of all, its beak size. The Grants were struck by the variation in beak sizes. To us, a fraction of an inch seems insignificant. But was it important to a finch? Did it make a difference in the food the finch ate? Did it matter to its survival?

To find out, the Grants and their students also measured finch food. Every morning they scoured the island with binoculars to see what the birds ate for breakfast. They got to know the food as well as they knew the birds, though it wasn't half as much fun. It required sifting through the dirt, counting the seeds, and measuring their size and hardness. They gave each seed a score based on how hard a bird had to struggle to crack it open.

By the end of their first year, the Grants had watched the ground finches on Daphne Major eat 4,000 meals. Medium ground finches have blunt beaks that are suited to crushing small seeds. All of the ground finches could be seen eating the small, soft seeds. But some of the birds—the ones with bigger beaks—could tackle the large, tough seeds of a plant called *Tribulus*. This plant is as tough as its seeds. It can withstand extreme conditions such as long droughts.

The Medium Ground Finch, *Geospiza fortis*
UNSM Angie Fox illustration.

These two male medium ground finches, *Geospiza fortis*, were on the same Galápagos Island at the same time. Notice the variation in beak depth.
Photo courtesy Peter Boag.

The Grants measure the size of finch beaks with great precision. Even small changes in beak size can influence the survival of the birds.
Photo courtesy Rosemary and Peter Grant.

The Grants and a graduate student, Peter Boag, returned to Daphne Major in the years following to measure the finches and their food. Nothing much changed. Then one year, a severe drought turned Daphne Major into the science experiment of a lifetime. The drought prevented many of the plants from producing seeds that year, so the finches were dependent on the seeds left over from the previous year. Once the finches ate all of the abundant small seeds that had been produced the year before, they were left with big tough seeds. Then a life-and-death struggle took place. When the Grants and Peter Boag returned to the island the next year, they were shocked by what they found.

Discover for yourselves whether variations in beak size can make a difference in finch survival and evolution. In this activity you'll meet some medium ground finches, measure their beaks, become a beak yourself and test your food-gathering skills, graph the finches and seeds on Daphne Major, and see whether changes in the environment can push the finches toward a different beak.

PART ONE
Islands & finches

The Galápagos are strange and wonderful islands. This string of volcanic mountaintops rise out of the Pacific Ocean floor about a thousand kilometers (600 miles) off the coast of South America at the equator. When these islands first emerged from the sea floor, they were simply lifeless piles of lava rocks. About three million years ago, new species began to arrive. The barren lava soils, equatorial heat, sparse rain, and isolation make this a very harsh world to survive in. Only some species have the equipment to live in these tough conditions, making it home to some unique plants and animals. These extreme conditions make the Galápagos very interesting to biologists.

Work with a partner

Each team will need:

- 6 Finch Beak sheets
  (10x larger
  than life size)
- Finch Beak
  Measurement Chart
- compass
- a ruler (cm & mm)

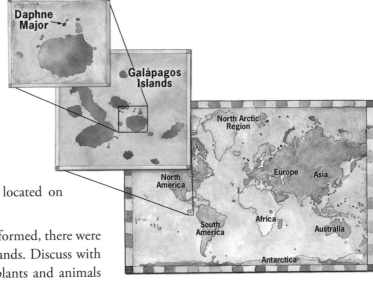

Lonnie Broden illustration.

1 Find the Galápagos

a Find where Daphne Major is located on each of the maps.

b When the Galápagos were first formed, there were no plants or animals on the islands. Discuss with your partner some ways that plants and animals from South America could have found their way to these rocky islands.

2 Measure Bird Beaks

No two animals are exactly alike. Differences between organisms of the same species in features like size, color, and abilities are called variations. Usually variations are slight and don't make much difference. Sometimes a variation will give one creature an advantage over another.

Rosemary and Peter Grant have visited the Galápagos every year for more than 30 years. They return to the island of Daphne Major to count the finches and band newly hatched birds. This puts them on a first-name basis with the finches that live on Daphne Major. The Grants pay attention to variations between each finch on the island. Learn how they measure finch beaks and discover some variations for yourself.

The Grants take precise measurements of their favorite finch, the medium ground finch, *Geospiza fortis*. You will be using a compass and ruler to measure beaks from pictures of medium ground finches that are 10 times larger than life.

a Collect your compass and ruler.

b How to measure beak depths: Measure the beak depth by putting it between the tips of a compass. Measure from the arrow at the top of the beak to the arrow just under the chin.

c Transfer the tips of the compass to the ruler without changing its opening. Measure in centimeters the space between the tips. If the compass tips have moved in the process, then re-measure the beak.

d Record the measurements for this bird on the Finch Beak Measurement Chart. You can identify each bird by its band number.

e Measure each of the six finch beaks and record your data on the chart. Then calculate the average beak size and write it on the chart.

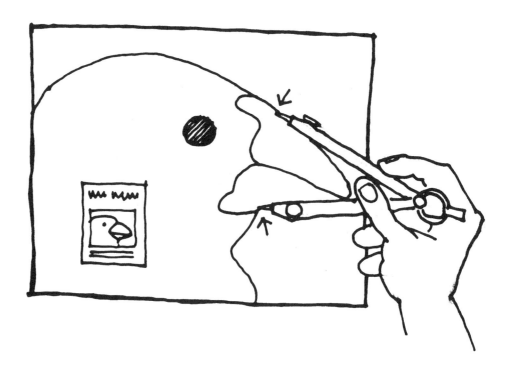

Galápagos Finches: Famous Beaks

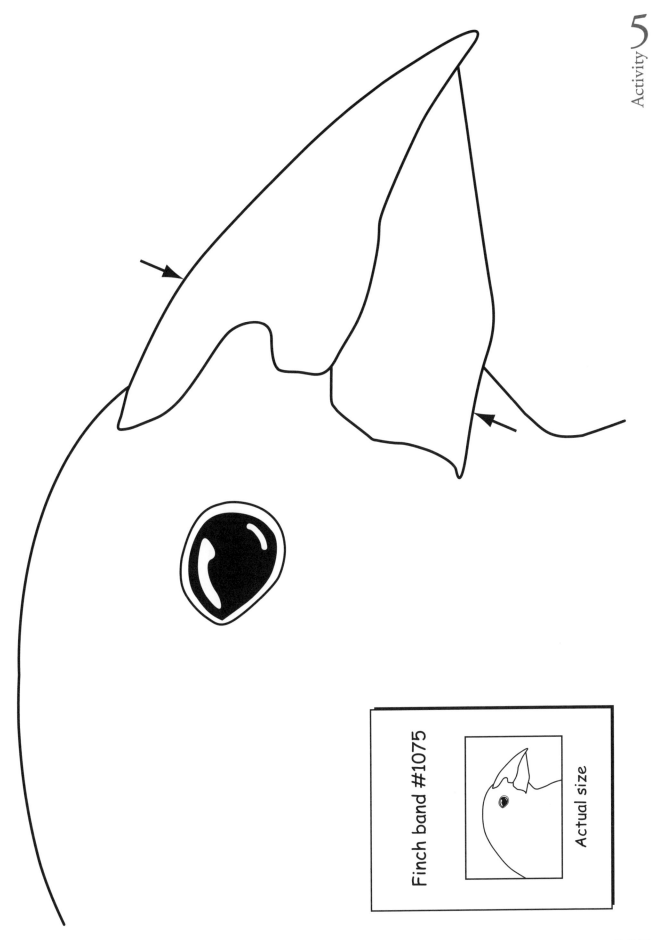

Finch band #1075

Actual size

Virus and the Whale: Exploring Evolution in Creatures Small and Large

Finch band #1999

Actual size

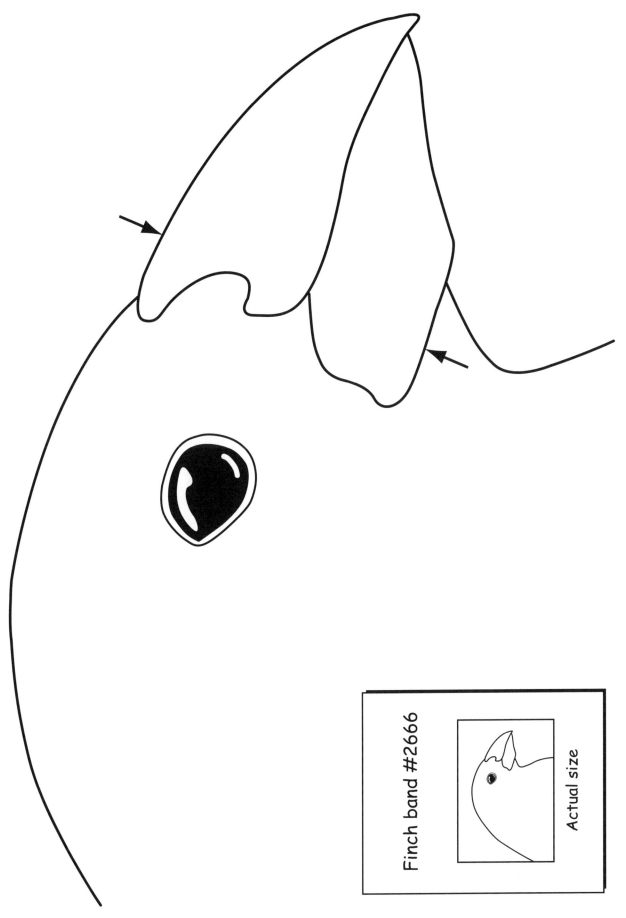

Finch band #2666

Actual size

Virus and the Whale: Exploring Evolution in Creatures Small and Large

Finch band #3527

Actual size

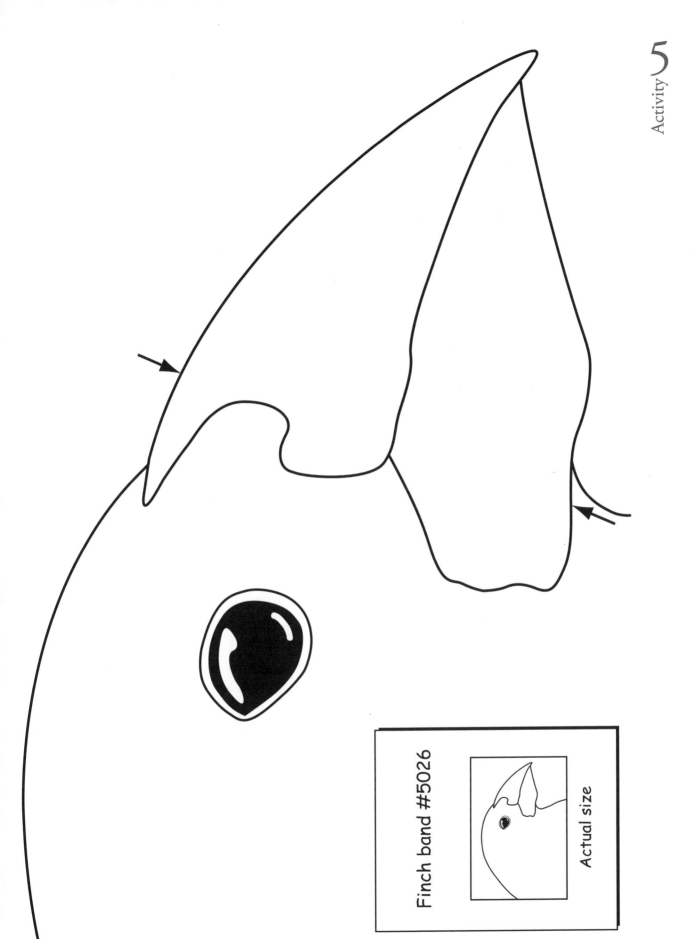

Finch band #5026

Actual size

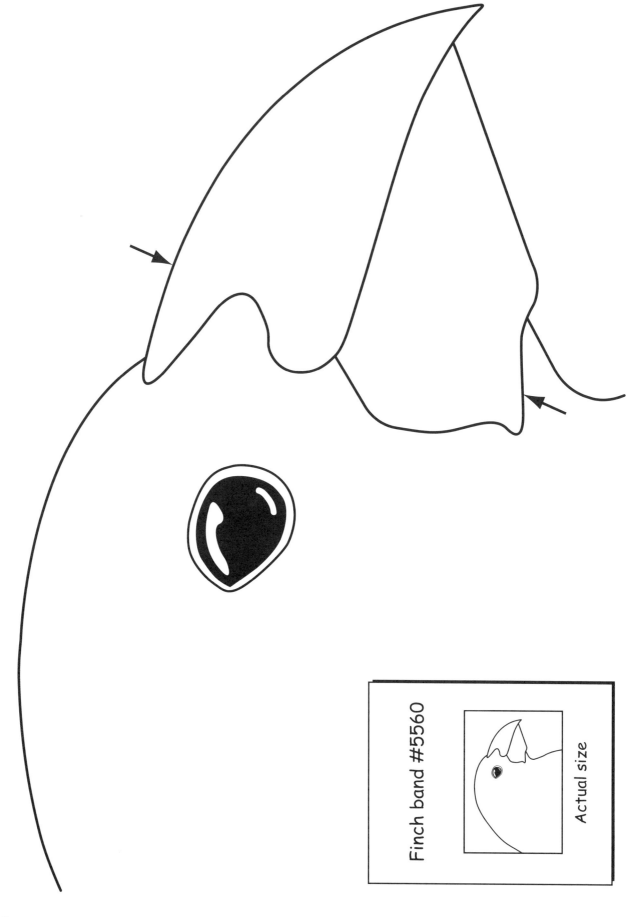

Finch band #5560

Actual size

Finch Beak Measurement Chart

MEDIUM GROUND FINCH BAND NUMBER	BEAK DEPTH IN CENTIMETERS	BEAK DEPTH IN MILLIMETERS
Finch 1075		
Finch 2666		
Finch 5560		
Finch 3527		
Finch 5026		
Finch 1999		

f The finch beaks you measured were enlarged 10 times. The actual range of beak variation measured by the Grants is in millimeters, not centimeters. We enlarged the finch pictures so the differences would be more visible to you. Now you need to multiply by 10 (this shows them in millimeters). Calculate the beak sizes in millimeters and insert them in the chart above.

g Now sum up your findings: How many different beak sizes did you find?

h What was the size in millimeters of the largest beak that you measured?

i What was the size in millimeters of the smallest beak that you measured?

j What is the difference in millimeters between the size of the largest and the smallest beaks that you measured?

4 Consider This

Do you think that tiny variations in beak size matter for survival? Why or why not?

PART TWO
Battle of the beaks

Medium ground finches eat a range of food including about two dozen kinds of seeds. These seeds range from small soft ones to seeds in hard shells that are tough to open. One of the toughest seeds on Daphne is from a plant called *Tribulus*. The seeds from *Tribulus* are about eight times harder to open than the soft seeds that finches also eat.

In North America *Tribulus* is a common weed that produces some of the toughest, meanest seeds around. Sometimes it is called puncture vine because the seeds have a talent for jamming themselves into bare flesh and bicycle tires.

Finches eat the easy food first. When soft seeds are plentiful the finches dine on those. When soft seeds are not available, the finches resort to the harder-to-open seeds. The scientists on Daphne Major observed that only finches with bigger beaks are able to crack open the tough seeds like *Tribulus* to get the food inside.

In this activity you will build two different types of beaks and test their food-nabbing effectiveness. Find out in the game of survival whether beak size matters.

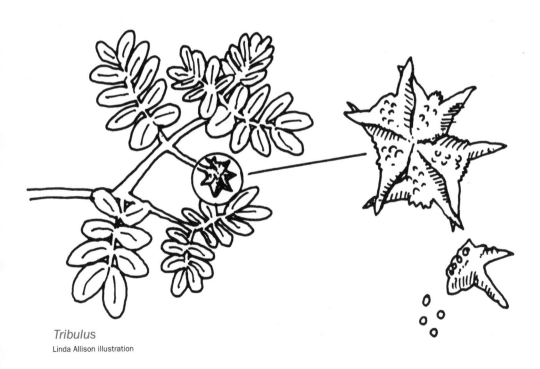

Tribulus
Linda Allison illustration

Work with a partner

Each team will need:

- Battle of the Beaks: Normal Year Chart
- Battle of the Beaks: Drought Year Chart
- beak materials: 4 popsicle sticks, 2 rubber bands, 9 pennies
- food supply: 1 teaspoon large seeds (e.g., garbanzos) and 1 teaspoon small seeds (e.g., mustard seeds)
- cm ruler
- tape
- sheet of paper ("stomach")
- paper plate ("island")
- timer

1 Build a Small Beak

Use popsicle sticks to test the difference between big beaks and small beaks.

a To build a small beak, you will need two popsicle sticks, a rubber band, and three pennies.

b Measure and mark a stick with a line 3 centimeters from the tip.

c Rubber band the two sticks together just past the 3 centimeter line.

d Make a stack of three pennies. Hold the stack together with a skinny piece of tape.

e Slide the stack of pennies between the two sticks. Position them just behind the rubber band.

f Press on the open ends to open the beak.

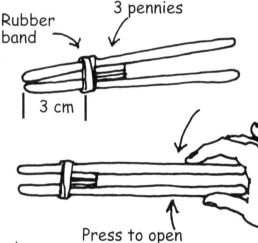

2 Build a Big Beak

a To build a big beak you need two popsicle sticks, a rubber band, and six pennies. Measure and mark a stick with a line 6 centimeters from the tip.

b Rubber band the two sticks together just past the 6 centimeter line.

c Make a stack of six pennies. Hold the stack together with a skinny piece of tape. Slide the stack of pennies between the two sticks. Position them just behind the rubber band.

d Press on the open ends to open the beak.

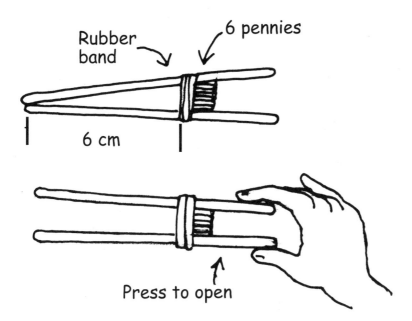

Rubber band

6 pennies

6 cm

Press to open

3 Beak Testing: A Normal Year

In this test your finch stays alive by gathering seeds on the "island" and collecting them in your "stomach." You will have one minute to gather up as much food as you can.

a Decide who will use the big beak and who will use the small beak. Write each name on the Battle of the Beaks: Normal Year Chart.

b Before you start, predict which beak will be the better tool for gathering seeds:

c In a normal year, there is a mix of large and small seeds. Prepare a seed supply for a normal year grabbing an equal pile of large and small seeds then scattering them on the island (the paper plate). Each beak needs a sheet of paper to act as a stomach for its seeds.

d Grab a few seeds and practice with your beak.

e Decide who will be the timer and who will be the beak tester.

f Ready to test? When the timer says go the tester will have one minute to use the beak to put as many seeds as possible in the stomach.

g When the minute is up, count the seeds that you gathered. Record the performance for your beak on the Battle of the Beaks: Normal Year Chart. Repeat the exercise for test #2.

Record data

"Saucer Island"   "Stomach"

Battle of the Beaks: Normal Year Chart

Beak Type	Normal Year (A Mix of Seeds)	
Small Beak Name of tester:	Test # 1	
	Test # 2	
	Total	
Big Beak Name of tester:	Test # 1	
	Test # 2	
	Total	

h Switch places and have your partner complete Test 1 and Test 2 for the other beak size.

i Add up the totals and record them on the Battle of the Beaks: Normal Year Chart. Which beak size gathered the most seeds?

4 Beak Testing: A Drought Year

One year no rain came to the island. Many plants failed to bloom and produce new seeds. All the medium ground finches ate the small, soft seeds first, leaving mostly large, tough seeds, so now big seeds dominate the menu.

a Prepare a seed supply for a drought year by leaving mostly large seeds and only two or three small seeds on the island.

b Repeat the steps you did for the Normal Year. You and your partner should complete Test 3 and 4 for both beak sizes. Record your data under Battle of the Beaks: Drought Year Chart.

Battle of the Beaks: Drought Year Chart

Beak Type	Drought Year (Mostly big, tough seeds)	
Small Beak Name of tester:	Test # 3	
	Test # 4	
	Total	
Big Beak Name of tester:	Test # 3	
	Test # 4	
	Total	

c Add up the totals and record them on the Drought Year Chart. Which beak size gathered the most seeds?

5 Review the Data

a Which beak size gathered the most seeds in a normal year?

b Which beak size gathered the most seeds in a drought year?

c Review your prediction under 3b (page 137) (Beak Testing: A Normal Year). Was your prediction correct or not?

d How does a change in environment (drought) affect which beak size gathers the most seeds?

6 Consider This

"Natural selection" occurs when the environment favors or selects some variations over others. You have tested two variations of beaks, large and small. In the drought environment, which beak variation is favored? Why?

PART THREE
Survival on Daphne Major

The scientists on Daphne Major observe everything on the island, and they keep a careful record of their data. In 1977 and 1978 they recorded a spell of over 500 days in which no rain fell. During this extremely dry period, many plants failed to produce seeds.

Investigate some measurements from the scientists' field notes. Turn their data into graphs to get a picture of what happened to the food supply and the finch population after the drought.

Work with a partner

Each team will need:

- Seed Abundance Graph
- Finch Population Graph

1 Tracking the Seed Supply

The observers on Daphne Major tracked seed abundance by first measuring a square meter area of ground and then sifting through the soil to count every seed. This was done at many different places to get an accurate count. They repeated the count every six months. Here are their data:

Field notebook seed count (measured in grams per square meter)

January 1976	**7.5**	January 1977	**8.0**	January 1978	**2.0**
July 1976	**10.5**	July 1977	**5.5**	July 1978	**3.5**

a Using the entries from the field notes, enter the data as a dot on the graph for each date. Connect the dots to complete the graph.

b Review the Seed Abundance Graph. During what month and year did the seed supply shrink to its lowest amount?

c During what month and year was the seed supply most abundant?

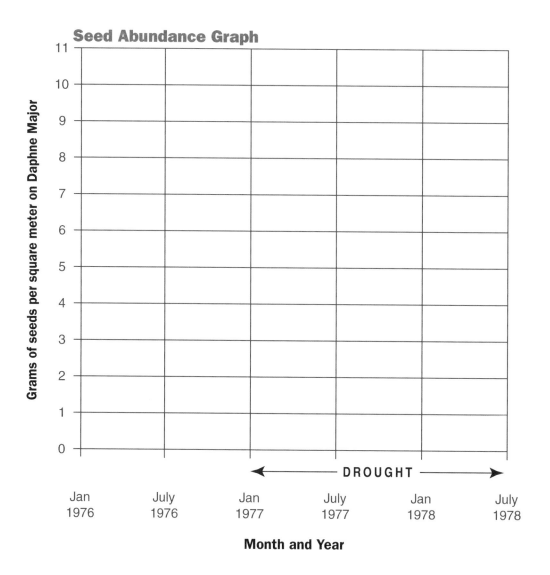

Seed Abundance Graph

2 Counting the Finch Population

The finches were counted every six months. Here are the data for the same period of time as the seeds were measured.

Field notebook finch count:

January 1976	**1100**	January 1977	**850**	January 1978	**200**
July 1976	**1400**	July 1977	**400**	July 1978	**357**

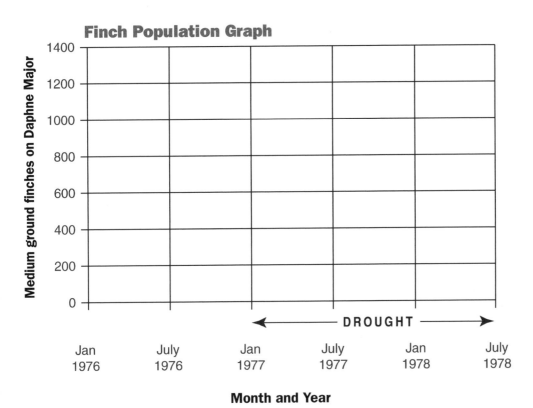

Finch Population Graph

a Using the entries from the field notes, enter the data as a dot on the graph for each date. Connect the dots to complete the graph.

b Review the Finch Population Graph. When was the finch population the lowest?

c When was the finch population the highest?

3 Think About Seeds and Finches Together

Compare the graphs side by side.

a What happened to the finch population when the seed supply shrank to its lowest amount? How do you account for this?

b When the seed supply increased, what happened to the finches? How do you account for this?

4 Bigger Beaks, But Why?

When the team returned to Daphne Major, they found only one in seven finches survived the drought. When they measured the survivors, they found that most were finches with big beaks. Why do you think bigger-beaked birds survived better than the smaller-beaked birds?

5 Consider This

Beak size is a variation that is passed from parents to offspring. When the new generation of young finches was measured in 1978, there were many more young birds with larger beaks. What happened?

PART FOUR
Be a science reporter

Write a short news story about the medium ground finches on the island of Daphne Major. Tell your readers about how the drought of 1977 led to changes in the characteristics of the finch population there. Based on what you have learned, explain why you think the finch population in the next generation had larger beaks after the drought.

P.S. Don't forget the headline.

HUMANS & CHIMPS:
All in the Family

No one would mistake you for a chimpanzee. Chimps have long arms and short legs, large canine teeth, a body covered with hair, a bent posture, and they walk on their legs and knuckles. Humans, on the other hand, have short arms and long legs, small canine teeth, relatively hairless bodies, and an upright posture. Take a trip to a zoo and check out the differences. Chimps resemble the gorillas, orangutans, and other apes behind bars more than they resemble the onlookers. As for humans, there is no one like us on Earth, right?

Wrong. Scientists were surprised to learn from DNA studies that humans are genetically very similar to chimps. The differences in our DNA are very hard to find. Our genes match so closely that we can catch many of the same diseases. Humans can even receive blood transfusions from chimps. If you think there is no one like us, think again.

Scientists are thinking twice, too. Svante Pääbo is the director of the Max Planck Institute for Evolutionary Anthropology in Germany. He and his colleague Henrik Kaessmann compare chimpanzees and humans to understand what their genetic similarities mean. If our DNA is so close, does it mean that chimps are our closest living relatives? If so, when did we share an ancestor, and how did we evolve to be different?

Two humans
Photo courtesy of SMM
Four chimps
Photo courtesy Curt Busse

Svante Pääbo is director of the Max Planck Institute of Evolutionary Anthropology in Germany.
Photo courtesy Svante Pääbo.

Henrik Kaessmann is at the University of Lucerne, Switzerland.
Photo courtesy Henrik Kaessmann.

DNA is the best tool we have for investigating how closely two species are related. Before DNA was discovered, scientists drew family trees based on similarities in anatomy, the physical structure and appearance of organisms. They relied on the fact that close relatives look more alike than unrelated individuals. In some cases, however, looks can be deceiving. Two friends may look as alike as sisters, while two sisters may look like they belong in different families. Sometimes it takes DNA "fingerprinting" to settle a question of family ties.

How do scientists read family histories in DNA? DNA is a long molecule made up of four chemical substances called nucleotides (NEW-klee-oh-tides). The nucleotides are named Adenine, Thymine, Cytosine, and Guanine, and they are usually abbreviated as A, T, C, and G. The nucleotides line up along the DNA molecule's length like words in a sentence. When an organism reproduces, DNA copies its nucleotides. The copying process is not exact, and sometimes a nucleotide may be left out, or two nucleotides may be switched. These changes are called mutations.

Our DNA is a combination of the DNA we inherit from our parents, plus the new mutations (changes in the nucleotides). When we produce children, we pass along our DNA plus new copying mistakes, and the mutations keep adding up. Most of these mutations have little or no effect on an organism, but their presence in our DNA helps to preserve a record of our accumulated changes over time. The mutations can be used like a trail of breadcrumbs to trace a species back to its ancestors.

Scientists line up the DNA sequences of different species, compare the nucleotides letter for letter, and count the differences. The more differences that accumulate

A guide to DNA

DNA is the molecule that stores the recipe needed to produce an organism.

DNA strands
DNA is made up of two backbones and their bases, which are known as **strands**. These strands can be pulled apart with high heat or harsh chemicals.

Backbone is made of sugars and phosphates. It forms the outside edges of the DNA molecule.

Bases are the "letters" in the genetic code. A base and the piece of backbone to which it is attached are known as a **nucleotide**. DNA contains four kinds of nucleotides:

Guanine (G)
Cytosine (C)
Adenine (A)
Thymine (T)

Nucleotides form complementary pairs in DNA. In DNA, **A** pairs only with **T**, and **C** pairs only with **G**.

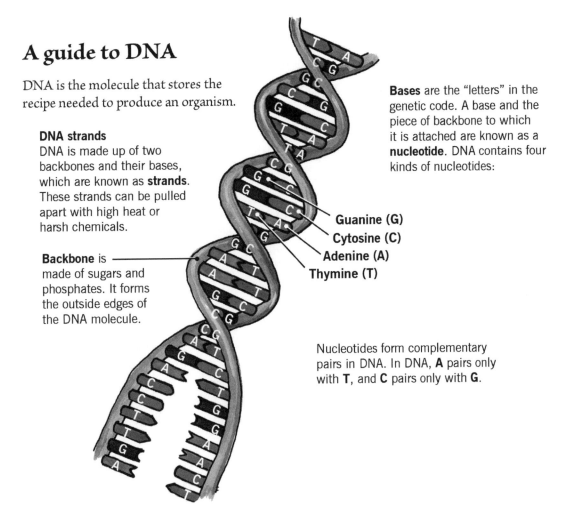

Science Museum of Minnesota (SMM) Adam Wiens, Lonnie Broden illustration.

between two species, the longer they have been evolving separately. DNA works as a kind of clock that ticks off evolutionary time.

Svante Pääbo has been counting the DNA differences in humans and other species for many years. Pääbo was the first scientist to develop a way to extract DNA from fossil humans and compare it to the DNA of living humans and chimps. Amazingly, he was able to extract the DNA from the arm bone of a 40,000-year-old Neanderthal, an extinct human-like species that lived in Europe. Then he compared it to samples of human and chimp DNA. Based on the number of differences, he estimated that Neanderthals and humans had a common ancestor half a million years ago. Chimps shared an ancestor with Neanderthals and humans much further back in time.

Pääbo and his colleague, Henrik Kaessmann, then traced the ancestry of humans and chimps to find out when they shared a common ancestor. The scientists took a segment of human DNA about 10,000 nucleotides long. The segment is called Xq13.3 for its location on the X chromosome. The scientists counted the number of nucleotides that were different. They discovered that humans and chimps averaged only about 100 differences out of 10,000 nucleotides on that length of DNA. That is a difference of about 1%.

Genetic studies like Pääbo's support a rich fossil record of human and chimp evolution. Researchers have uncovered the bones of ancient primates that shared ape and human characteristics. From these fossils, a general picture of our most recent common ancestor is emerging. The great, great, great (etc.) grandmother of chimps and humans was a primate who lived in Africa about five or six million years ago.

One of the most famous fossil discoveries so far is Lucy, an almost complete skeleton of a female ancestor who lived about four million years ago. Lucy stood erect and walked like a human, but she was as small as a chimpanzee, and she had a chimp-like brain. Lucy shows us just how chimp-like our ancestors were a few million years back.

Svante Pääbo and other DNA researchers are now focusing on the 1% difference between human and chimp DNA. They are hoping to find answers to the question everyone is asking: Why do we look so different when our genes are so much alike?

In this activity you are invited to compare humans and chimpanzees, inside and out, learn how to tell time with DNA, and decide for yourselves whether we should keep chimps in a zoo or invite them to Thanksgiving dinner.

PART ONE
Chimps vs. humans

People have long been fascinated by how similar chimpanzees look to humans. But just how similar—and different—are we? Take a close look and compare a chimp and a human side by side. Create a set of human and chimp attribute cards based on what you find.

Work with a partner or in a small group

Each team of two or more will need:

- Chimpanzee vs. Human sheet
- Comparing Creature Features sheet
- Feature Cards (blanks; one sheet per group)
- Hidden Feature Cards (one sheet per group)
- tape
- scissors

1 What's Human… What's Not.

a Check out the Chimpanzee vs. Human sheet. Cut out the Comparing Creature Features sheet and match the half to the other species on the Chimpanzee vs. Human sheet. Tape the half into place if it helps.

Chimpanzee vs. Human

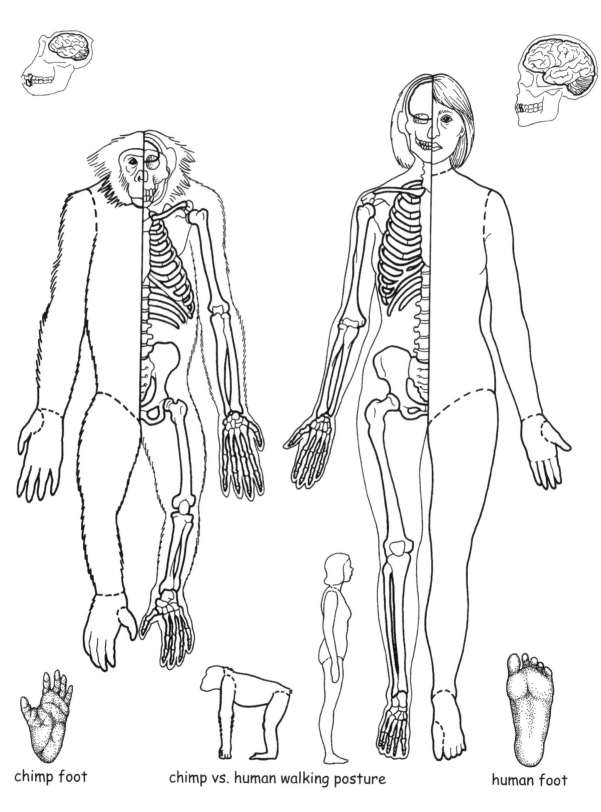

chimp foot

chimp vs. human walking posture

human foot

Page 4-36 Comparative Anatomy and page 5-16 Footprints and Foot Bones from THE HUMAN EVOLUTION COLORING BOOK by ADRIENNE ZIHLMAN, Copyright (c) 1982 by Coloring Concepts, Inc. Reprinted by permission of HarperCollins Publishers Inc.

Humans & Chimps: All in the Family

Comparing Creature Features

Cut out the half drawings along the dotted lines. Match each half to the opposite species on the Chimpanzee vs. Human sheet for easy comparison.

✂ Cut on dashed lines.

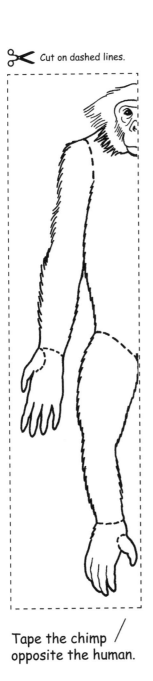

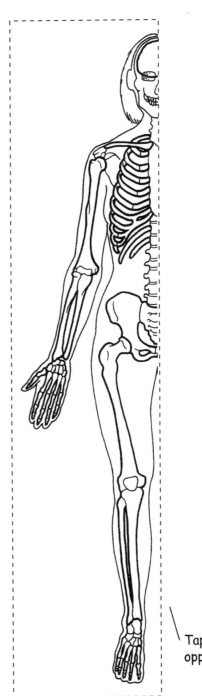

Tape the chimp
opposite the human.

Tape the human
opposite the chimp.

Virus and the Whale: Exploring Evolution in Creatures Small and Large

Fill them with your observations.

Feature Cards

Cut cards apart on the dashed lines.

Both chimps and humans have...

Humans have...

Chimps have...

b Write your observations about the similarities and differences on the blank Feature Cards. First cut the cards out. Then take about ten minutes and list as many features as you can. Then write one observation per card.

For instance:

"Chimps have fur."

" Humans have smooth skin with tiny hairs."

Here are some things to notice:

Bone shapes	Finger lengths	Teeth
Ear position	Skull shape	Limb lengths
Posture	Butt muscles	Face shapes

2 Sorting Out Differences

a Sort all the Feature Cards you wrote into three piles: Chimps, Humans, or Chimps and Humans.

b Now cut and read the Hidden Feature Cards. These cards are about human and chimp features that you can't see in a picture. Sort them into the same three piles. You may have to guess where they fit. *(Don't look now, but the answers are on the last page of Activity 6.)*

3 Consider This

Would you say that humans and chimps are more different than they are alike… or more alike than they are different? Why?

Hidden Feature Cards

✂ Cut cards apart on the dashed lines.

Uses hand gestures and facial expressions to communicate.

They are also able to use a computuer keyboard to communicate.

Has a long childhood and loves to play.

Develops a lifelong bond with its mother.

These are social animals that live in communities. The top ranking male leads the community.

These animals yawn (although no one knows exactly why). Yawning is contagious among group members.

Has a large brain relative to body size.

Males weigh up to 110 pounds and females up to 85 pounds. The average lifespan is 50 years.*

Lives in territories that are defended by roving males. Practices aggressive and sometimes lethal behavior against its neighbors.

When they are sick they dose themselves with medicinal herbs.

Linda Allison illustration

PART TWO
It's molecular time

DNA is no ordinary stuff. DNA carries the recipe for assembling proteins into living organisms. This long molecule also acts like a clock, helping scientists estimate how long two species have been separated from a common ancestor. When DNA makes a copy of itself, it isn't always perfect. Mistakes can happen. One nucleotide (Adenine, Thymine, Cytosine, or Guanine) might be missing, duplicated, or two nucleotides might switch positions. Scientists like Svante Pääbo have discovered that these mistakes (mutations) accumulate at a regular rate over millions of years, like the steady tick of a clock. Scientists can use this knowledge to date different copies or generations of DNA. This is what scientists call the "molecular clock."

See if you can tell time using the molecular clock method. First, compare copies of tiny segments of DNA for differences (mutations) in the nucleotides. Then use the mutation rate to date the copies.

Work with a partner

Each team will need:

- Generations of Copies sheet
- scissors

1 Down to DNA

a Below is a short segment of DNA that is nine nucleotides long. The first line of nucleotides is from a living organism. The second line is from a close ancestor. Underline the nucleotide in the ancestor DNA line that is different from the living DNA. This difference in the DNA code is a mutation site.

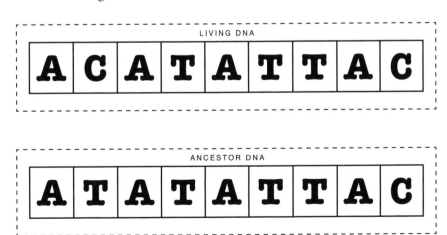

LIVING DNA

A C A T A T T A C

ANCESTOR DNA

A T A T A T T A C

The letters represent nucleotides:

A = Adenine

T = Thymine

C = Cytosine

G = Guanine

2 Sorting Out Sequences

a Cut out all the sequences of DNA from the Generations of Copies sheet.

b You have a nine nucleotide sequence from a section of DNA similar to what you might find from a living human. Look for the sequence with the label "living DNA."

c Now find the closest ancestor. It is the sequence that differs by only one nucleotide. The remaining sequences you have cut out are each older ancestors. Each older ancestor has more mutations or differences.

d Sort the sequences in order from the living human (the present) to the oldest ancestor (longer and longer ago).

3 Consider This

Now make the DNA tell time. Assume that the rate of mutation in this DNA segment is one difference for every 10,000 years.

a Based on the rate of mutation, how many years different are the oldest and the newest DNA segments?

b How long ago did the oldest ancestor live?

Generations of Copies

Below is a sequence of nucleotides from DNA that is similar to what you might find from a living human ("Living DNA"). The rest are sequences from five different ancestors. Cut out the sequences and sort the sequences starting with the living human to the most distant ancestor.

LIVING DNA

A C A T A T T A C

A T A T A T T A C

T T A G T T A A A

A T A G A T T A C

C A T T A G T T A

A T A G T T A A

PART THREE
Mutations up close

Scientists like Pääbo line up the DNA sequences of different species. They compare the nucleotides A, T, C, G, letter for letter, and count the differences. The differences are the number of mutations in the DNA code that have accumulated over time. The more differences that accumulate between species, the longer the species have been evolving separately.

The genetic code of humans and chimps is billions of letters long. Working out a method for comparing DNA sequences between the two species is one of the problems that genetic scientists must solve. Today this is your challenge too. Only about 1% of the DNA in the chimp and human genes is different. Can you pinpoint the differences?

Work with a partner

Each team will need:

- Chimp vs. Human DNA Sequences sheets, pages 162 and 163, parts 1 and 2 (taped together)
- scissors

1 Comparing DNA

    a Compare the Chimp vs. Human DNA Sequences. The sequences are located on the X chromosome, and they are called Xq13.3. These are the small sections of the DNA that Svante Pääbo and his group use to make chimp/human comparisons. Look for any differences (mutations) between the chimp and human sequences.

    b How to read the chart:

- In the chart you will find the same stretch of DNA (about 2,700 nucleotides long) for a chimpanzee (top) and a human (bottom).

- The vertical lines show where the DNA is the same for chimps and humans.

- A tiny Pääbo figure shows where there's a difference in the chimp and human DNA.

    c How many Pääbos can you find?

2 Consider This

Did you expect more or fewer differences between chimp and human DNA? Why?

Part 1

The sequences are arranged in paired rows, each labeled CHIMP and HUMAN, beginning with CHIMP DNA (START) and HUMAN DNA (START).

DNA sequence courtesy of Henrik Kaessmann and Svante Pääbo.

✂ Cut and tape to the top of DNA part 2.

Humans & Chimps: All in the Family

Part 2

Line up the columns of nucleotides on DNA part 1 with DNA part 2. Then tape the bottom of DNA part 1 to this line.

```
CHIMP T T A T G T A G G A A G A C C A A C A G T T G A G G T T T G C C T G T G A C T G A G G G T T T T T T G G G A C A T A G T A T T T T C A G T G C T A A A CHIMP
HUMAN G T A T G T A G G A A G A C C A A C A G T T G A G G T T T G C C T G T G A C T G A G G G T T T T T T G G G A C A T A G T A T T T T C A G T G C T A A A HUMAN

CHIMP A A T T G A G C A A T T T T G A A G A C T G C T T G A G G A G T C A A A G A T G A T T T C A T T T A T A A A A C A T G A A A T G T T A T A A A A A A G CHIMP
HUMAN A A T T G A G C A A T T T T G A A G A C T G C T T G A G G A G T C A A A G A T G A T T T C A T T T A T A A A A C A T G A A A T G A T A T A A A A A A G HUMAN

CHIMP A T C C A T T A A C C T A T T T C T T A T T C C A A T A T T T T T C A A T T T T G T A C T A T C T G C C A A C C C T C A T C A A T T C C T A A C T C T CHIMP
HUMAN A T C C A T T A A C C T A T T T C T T A T T C C A G T A T T T T T C A A T T T T G T A C T A T C T G C C A A C C C T C A T C A A T T C C T A A C T C T HUMAN

CHIMP T T C C A G A A A C C T T C A A T A T C T A A A T G C T T C C A T T T C T C C T A G T C T C A C A T A G C A T C C T C C C T C A T C T T A T T T C C T CHIMP
HUMAN T T C C A G A A A C C T T C A A T A T C T A A A T G C T T C C A T T T C T C C T A G T C T C A C A T A G C A T C C T C C C T C A T C T T A T T T C C T HUMAN

CHIMP C A T T C T C C T G C T G C A C C A G C C T G G C A A G G C C C T A A T C G C A G G T T A A G T C C A C A A C A G G T T T T C T T T C T T A C T A T C CHIMP
HUMAN C A T T C T C C T G C T A C A C C A G C C T G G C A A G G C C C T A A T C G C A G G T T A A G T C C A C A A C A G G T T T T C T T T C T T A C T A T C HUMAN

CHIMP C T A A A A T T A T T T T C C T C T A C C C T C T T T T T A C A T G A A C C T G A T T A G T A T C T T C T C C T A T T C C C A T G G T G G C T A A T C CHIMP
HUMAN C T A A A A T T A T T T T C C T C T A C C C T C T T T T T G C A T G A A C C T G A T T A G T A T C T T C T C C T A T T C C C A T G G T G G C T A A T C HUMAN

CHIMP T T G T T T T A G C C C T C A G C T C A A A G A G T A A A T T G A G A T T A T T A G G T G A G A A G A A A T C A A C T T C A T C T A C T G T C C T T T T CHIMP
HUMAN T T G T T T T A G C C C T C A G C T C A A A G A G T A A A T T G A G A T T A T T A G G T G A G A A G A A A T C A A C T T C A T C T A T T G T C C T T T T HUMAN

CHIMP T G A A T T A T C A C A C T C T T C T A A C C T C C C T C A G G T T T C T C T T T C A T T G A T T G T T C C C T T T C A C T C C T A G A A A A T C A T CHIMP
HUMAN T G A A T T A T C A C A C T C T T C T A A C C T C C C T C A G G A T T C T C T T T C A T T G A T T G T T C C C T T T C A C T C C T A G A A A A T C A T HUMAN

CHIMP C T T A T T C T A G C T C T T T G T T G C C T T C C T G C C C T T C T C T C C A T A G T T G T T T A C A C C A G C T T T T C C T C A C T T T A T A T C CHIMP
HUMAN C T T A T T C T A G C T C T T T G T T G C C T T C C T G C C C T T C T C T C C A T A G T T G T T T A C A C C A G C T T T T C C T C A C T T T A T A T C HUMAN

CHIMP T T A C C C A G A A A G A G G T G A G A A T A G A A T T C C C A G T G C C T C C A A T G C C A G C C T T A C A G A A G T A A T T A C T C T T G G T A C CHIMP
HUMAN T T A C C C A G A A A G A G G T G A G A A T A G A A T T C C C A G T G C C T C C A A T G C C A G C C T T A C A G A A G T A A T T A C T C T T G G T A C HUMAN

CHIMP T C A C T T G A A A G A A G G G T A T G T T G G T G A G T G C C A A A T C C T G A A G G G A A T G A C C A G C A T A G A G T T T T A A C C T A A T A A CHIMP
HUMAN T C A C T T G A A A G A A G G G T A T G T T G G T G A G T G C C A A A T C C T G A A G G G A A T G A C C A G C A T A G A G T T T T A A C C T A A T A A HUMAN

CHIMP G C C A T A A A A A G A A T G A G A T C A T T T C C T T T G C A G G G A C A T G G A T G G A G T T G G A A G C C A T T A T C C T T A G C A A T C T A A CHIMP
HUMAN G C C A T A A A A A G A A T G A G A T C A T T T C C T T T G C A G G G A C A T G G A T G G A G T T G G A A G C C A T T A T C C T T A G C A A T C T A A HUMAN

CHIMP T G G T G A G A A A C A A G A A A C A T T G G A G T C T G T T G G G A G C T G T T G G G T G A G A G A G A G C A T C A G G A A G A A T A G C T A A T G CHIMP
HUMAN T G G T G A G A A A C A A G A A A C A T T G G A G T C T G T T G G G A G C T G T T G G G T G A G A G A G A G C A T C A G G A A G A A T A G C T A A T G HUMAN

CHIMP A A C C C T G C A C A T C C T G C A C A T G T A C C C C T A A A C T T A A A A T A A A A G T T G A A G A A A A T A T G G CHIMP DNA (END)
HUMAN A A C C C T G C A C A T C C T G C A C A T G T A C C C C T A A A C T T A A A A T A A A A G T T G A A G A A A A T A T G G HUMAN DNA (END)
```

DNA sequence courtesy of Henrik Kaessmann and Svante Pääbo.

Activity 6

PART FOUR
Be a science reporter

Write a short news story about humans and chimpanzees. Tell your readers about how new DNA studies of humans and chimpanzees suggest they are close relatives. Based on what you have learned, explain how you think a modern chimpanzee and a modern human could have a common ancestor.

P.S. Don't forget the headline.

Answer: The hidden feature cards describe both humans and chimps, except for the card with the asterisk. This card describes the chimps.

WHALES:
Walking Into the Past

Whales with knees and toes? Incredible as it seems, whales once walked on legs and lived on land. Millions of years of biological change have erased the whale's legs from its body. But a faint trace remains. Hidden inside the streamlined body of many modern whales are tiny hip and leg bones. The story of how the whale—marvel of the oceans—evolved from a four-legged mammal is an amazing one. So far, it provides one of the best examples of how organisms change over time. And like whales, the story is still evolving.

Just about everything to do with whales was once a big puzzle, even what kind of animal it was. Fish or mammal? Scientists had trouble deciding. From the small dolphin to the enormous grey whale, these animals look and live like fish. They can't survive out of water—but on the other hand, they can drown in water. Every 15 minutes or so they have to swim to the water's surface to breathe. Like all mammals, whales are warm-blooded, give birth to babies rather than lay eggs, and nurse their young with milk. They even have a belly button.

Philip Gingerich, paleontologist at the University of Michigan, digs for fossil whales.
Photo courtesy University of Michigan Museum of Paleontology.

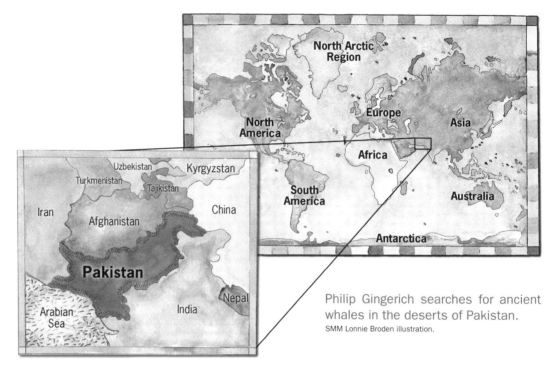

Philip Gingerich searches for ancient whales in the deserts of Pakistan.
SMM Lonnie Broden illustration.

So, if whales are mammals, how did they come to live like fish? Could their ancestors have been land mammals that gradually took to the oceans? When scientists first asked these questions, there was little fossil evidence to provide answers. Until, that is, paleontologist Philip Gingerich discovered a bone in the desert of Pakistan.

Many of the great scientific discoveries happen by accident, if the scientist is open to surprises. Philip Gingerich wasn't looking for whales when he arrived in Pakistan in 1977 with a team of international scientists. Dr. Gingerich was an expert on extinct land mammals. So he was disappointed to find that the first place he had targeted to look for fossils was an ancient seashore. The rocks were filled with fossilized snails and other shellfish—not the best place to find the bones of a land mammal.

Like kids in a giant sandbox, Gingerich and his team went to work anyway. After a week of scouring the exposed rocks, they found a few pieces of bone that looked promising: They could make out part of a pelvis and a backbone. Gingerich recalls joking with his team about finding a "walking whale" with hips and legs, but back then it was just a joke. The team assumed that the fossils were bits and pieces of an ancient elephant ancestor, a land mammal that had drifted out to sea after death.

But on a December day in 1979, the team found a specimen that was not so easy to explain. Embedded in rock as hard as cement, it was a curious fossil skull. Gingerich took the skull home to his lab. When the fossil was cleaned up, Gingerich could see that it was no bigger than a wolf's skull and had many wolf-like features. The teeth were a mix of canines and molars, and the nostrils were set close to the snout. But attached to the skull was a set of tiny thickened ear bones. The only animals on Earth, living or extinct, that have ear bones thickened this way are the whales.

Gingerich named the creature *Pakicetus* (pack-eh-SEA-tus), Whale of Pakistan. It lived about 48.5 million years ago. Based on the land mammal-like features of the skull, Gingerich hypothesized that this whale probably went into the water to feed but came out on land to rest and to give birth. After the unexpected discovery of *Pakicetus*, Gingerich began to hunt for whales in earnest. Whale fossils are large, and they are relatively easy to find in the desert. Sometimes the wind lends a hand by scouring away the sands and shales, leaving entire skeletons exposed. One valley in Egypt was especially rich in whales. On a series of visits starting in 1983, Gingerich and his team found virtually a whale a day. The final tally came to 379 whales. He found some with hipbones, but Gingerich also hoped to find legs and feet.

In 1989 Gingerich was mapping the spine of a well-preserved 50-foot-long skeleton of *Basilosaurus* (bah-sill-oh-SOAR-us), a fossil whale that lived about 10 million years after *Pakicetus*. Two-thirds of the way down the spine he found a small round bone standing vertically. This seemed out of place, and the top was weathered away. When the rest was excavated, the bone proved to be an upper leg bone with the knee joint preserved. No one had ever seen the knee of a whale before.

Counting down the spine, this bone was near the 48th vertebra. The whole team went to work brushing sand away from this part of all the *Basilosaurus* that had been mapped so far. Soon they found, astonishingly, not only more hips and upper legs, but also lower leg bones, anklebones, and finally, one by one, the bones of three tiny toes. These were the first hind limbs and feet ever found with a fossil whale skeleton.

Basilosaurus was a huge ocean-going whale, with retracted nostrils forming a blowhole halfway up its four-foot-long skull. Its body was equipped with legs, but they were too small to support the animal's weight on land.

Gingerich predicted that scientists would unearth many more missing links between land and sea whales. He wasn't disappointed. Some of the most important finds were yet to come. In the 1990s J. G. M. Thewissen and Gingerich discovered two whales that were almost as old as *Pakicetus*. Both had legs larger than *Basilosaurus*, even though the skeletons were only 10 to 12 feet long. One was named *Ambulocetus natans* (am-bue-low-SEA-tus-NAY-tans) and the other *Rodhocetus kasranii* (row-deh-SEA-tus-kaz-RAN-nee-eye). Like *Pakicetus*, both of these whales found their food in water and were good swimmers, but both still hitched their way ashore to rest and to give birth.

Some crucial parts of the *Rodhocetus* skeleton were still missing. The hands, feet, and tail were poorly known. But Gingerich persevered, and eventually his team found

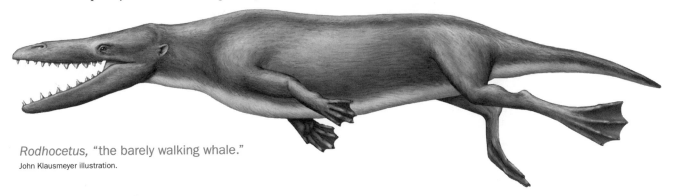

Rodhocetus, "the barely walking whale."
John Klausmeyer illustration.

a complementary *Rodhocetus* skeleton with hands and feet intact. He named this one *Rodhocetus balochistanensis* (bah-low-CHIS-stan-en-sis), after the province in Pakistan where it was found. This discovery brought new surprises. The middle three fingers of each hand retained a tiny hoof. And the anklebones proved to be the kind that belonged only to hoofed mammals known as artiodactyls (arty-oh-DAK-tils). Artiodactyls include cows, goats, pigs, and hippos. *Rodhocetus* combined features of an aquatic whale with features of a hoofed mammal all in the same skeleton.

Scientists doing DNA studies had already claimed that the whale's closest living relatives were artiodactyls like the hippopotamus, and here was confirmation. The fossil record and DNA evidence were now saying the same thing. *Rodhocetus* was like an arrow pointing backward to a hoofed ancestor and forward to an ocean-dwelling dolphin.

In this activity you'll have a chance to meet *Rodhocetus,* the barely walking whale, excavate and sort some "fossils" of your own, and learn how to read bones like a paleontologist to understand the whale's remarkable transition from land to sea.

PART ONE
Sweet excavation

Discovering fossils is not like in the movies where whole skeletons of ancient critters magically reveal themselves in all their magnificence at the flick of a whiskbroom. The process is more like reassembling bits of roadkill that were buried millions of years ago and cemented into rock. Finding fossil bones is just the first step. The bones then have to be carefully dug up, and then packed and shipped back to the lab. There they are cleaned, studied, and assembled. It is slow work. Try your own hand at excavating some fossil "bones" and making sense of your finds.

An exciting new discovery of "fossil pretzel bones" has arrived in your lab. Your job is to excavate, sort, and assemble the bones. Then see if you can assemble the parts well enough to identify what kind of pretzel creature these bones belong to.

Work with a partner

Each team will need:

- Pretzel Species sheet
- stiff paper or cardboard
- liquid glue

Either of two options:

1 Fossil Evidence Envelope
   - pretzel broken into 3–4 pieces
   - envelope

2 Cookie Fossil Rocks
   - cookie dough baked with 3–4 pretzel pieces inside
   - big nail or metal nail file

PREP NOTE: This activity can be done in either of the following two ways, depending on the amount of preparation time available:

Fossil Evidence Envelope:

To prepare, break a pretzel into about 3–4 parts and place the parts into an envelope. For more challenge, add extra pretzel parts, remove some, or try working with several shapes of pretzels and mix the parts.

Cookie Fossil Rocks:

To make each cookie fossil rock, mix pretzel pieces with enough cookie dough to contain them. Bake according to cookie directions. Hard, dry cookies will be difficult to excavate.

---

1 Making Sense of Fossils

a Look at the Pretzel Species sheet of some known species of the pretzel family to get familiar with their body forms. Open your "fossil evidence envelope" or excavate your "cookie fossil rocks" with a tool like a metal nail file. Lay out the pretzel parts.

b You may or may not have parts for a whole animal, but what you do have represents ONE species. Some of the parts may have been eaten by scavengers or gotten lost. Assemble the parts in a way that makes sense, using the shapes on the Pretzel Species sheet to help you sort and assemble.

c When you have assembled the parts of the pretzel animal, glue them to a sheet of paper.

d If you don't have a whole animal, draw in what you think are the missing parts.

2 Consider This

Does your pretzel look like one of the known species or is it a new variety? Describe how your fossil finds compare to the ones on the Pretzel Species sheet.

When scientists discover a new kind of animal, they have the honor of naming the new species. If you think your pretzel bones belong to a new and different pretzel species, make up a name, and write it here.

Pretzel Species: These are some members of the Pretzel Family

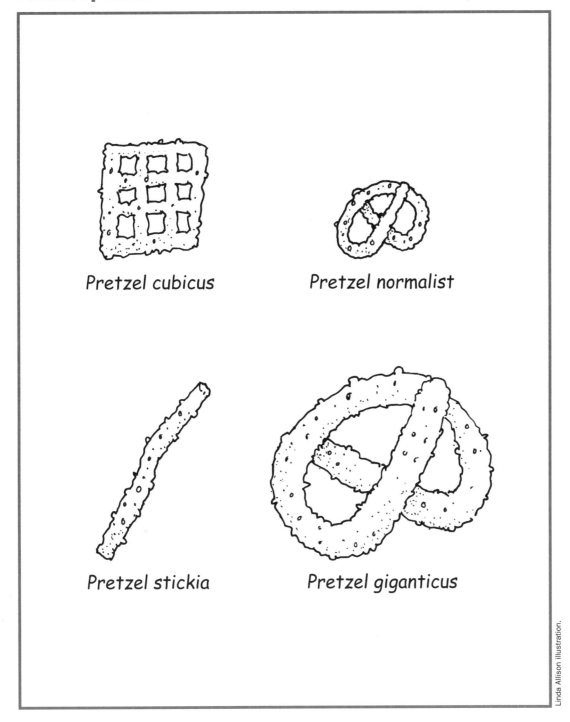

Pretzel cubicus

Pretzel normalist

Pretzel stickia

Pretzel giganticus

Linda Allison illustration.

PART TWO
Read some bones

When Philip Gingerich found the mystery bones in Pakistan, he knew they were not the bones of a modern animal. Gingerich knew he was digging in a layer of rocks that was about 47 million years old, which gave him a good estimate of about how old the bones were. He also knew from studying the rocks that the bones had been deposited near a shallow sea. He took the bones back to his lab and after many months of painstaking work, his team cleaned and assembled the skeleton of the animal he named *Rodhocetus*.

Even if a bone was buried for millions of years, a paleontologist can study a fossil bone and learn many things about the animal it came from. By comparing the shapes of fossil bones to modern bones, a lot can be learned about how the fossil animal might have lived and worked.

Work with a partner
Each team will need:
- One Mystery Fossil Bones card (cut separate cards for Fore Limbs, Hind Limbs, Skull & Teeth, OR Spine, Neck, & Ribs)
- One matching Some Known Bones sheet (Fore Limbs, Hind Limbs, Skull & Teeth, OR Spine, Neck, & Ribs)
- One Sum It Up sheet (Fore Limbs & Hind Limbs OR Skull & Spine)
- Scissors

1 Bone Reading
Study some pictures of mystery fossil bones excavated by Dr. Gingerich in Pakistan. See how much information you can "read" from these mystery fossil bones by comparing them to the bones of some modern animals alive today.

Every team should get a Mystery Fossil Bone card and the matching Some Known Bones sheet. Complete the questions in the challenge section of the Some Known Bones sheet.

MYSTERY FOSSIL BONES: SPINE, NECK, & RIBS

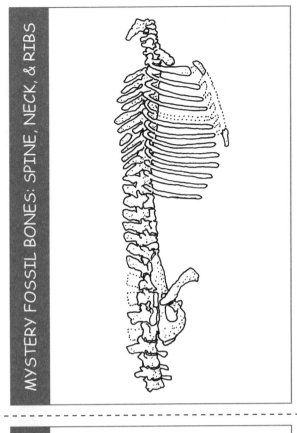

MYSTERY FOSSIL BONES: FORE LIMBS

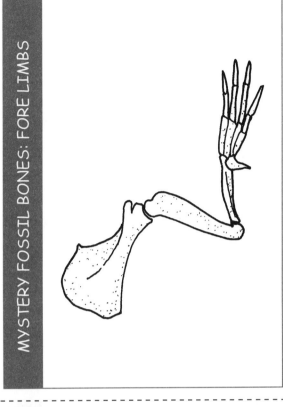

MYSTERY FOSSIL BONES: SKULL & TEETH

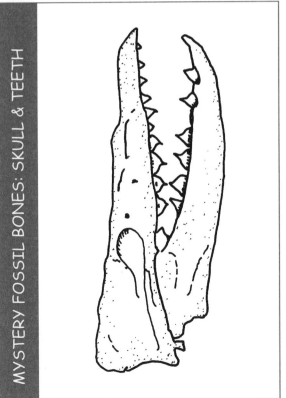

MYSTERY FOSSIL BONES: HIND LIMBS

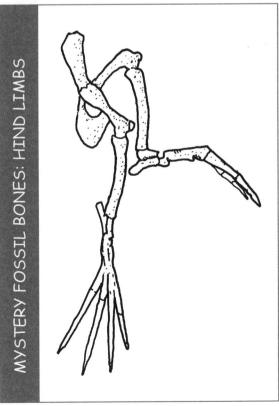

Virus and the Whale: Exploring Evolution in Creatures Small and Large

SKULL & TEETH: Some Known Bones

CHALLENGE:

FIND THE MODERN SKULL AND TEETH MOST LIKE THE MYSTERY FOSSIL BONES.

1. Get the card "Mystery Fossil Bones: Skull and Teeth."

2. Compare the mystery skull and teeth to the bones of modern animals shown on this page.

3. Which skull and teeth do you think are most like the Mystery Fossil Bones?

4. List the clues that helped you decide:

_____

_____

_____

_____

5. Based on the bone clues, what type of critter did the Mystery Fossil Bones belong to?

_____

_____

_____

An animal's skull acts like a hard hat to protect its brain. Sharp teeth are suited for tearing into flesh; flat teeth are better for grinding plant foods. Animals with large eye sockets pointing ahead have good forward vision while eyes to the sides are better for seeing all around.

● Wolves use their forward-pointing eyes and nose for hunting. Sharp teeth stab and hold prey.
● Monkeys use their keen eyes and nose for finding food. They use their diverse teeth to eat a mixed diet.
● Dolphins breathe at the water's surface from nostrils on top of their head.

Note how the shapes of the bones give clues to an animal's lifestyle.

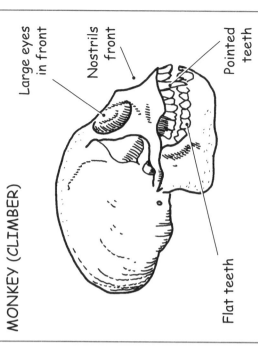

WOLF (RUNNER)

Large eyes in front

Nostrils front

Skull ridge anchors strong jaw muscles

Bone-cracking molars

Sharp teeth tear meat

MONKEY (CLIMBER)

Large eyes in front

Nostrils front

Pointed teeth

Flat teeth

DOLPHIN (SWIMMER)

Nostrils on top of skull

Eyes positioned on the sides

Pegged teeth grip fish

Linda Allison illustration

FORE LIMBS: Some Known Bones

CHALLENGE:

FIND THE MODERN FORE LIMBS MOST LIKE THE MYSTERY FOSSIL BONES.

1. Get the card "Mystery Fossil Bones: Fore Limbs."

2. Compare the mystery fore limbs to the bones of modern animals shown on this page.

3. Which fore limbs do you think are most like the Mystery Fossil Bones?

4. List the clues that helped you decide:

_____

_____

_____

_____

_____

5. Based on the bone clues, what type of critter did the Mystery Fossil Bones belong to?

_____

_____

_____

_____

Animals often use their fore limbs or front legs for specialized jobs like digging, paddling, and grabbing. Their back legs are mainly used for locomotion. Notice how different animals use their fore limbs in different ways:

- Wolves run on their long legs and compact feet. Strong shoulders support their body weight.
- Monkeys swing through trees with their long arms and grasping hands.
- Dolphins swim using their flippers as paddles.

Note how the shapes of the bones give clues to an animal's lifestyle.

WOLF (RUNNER)

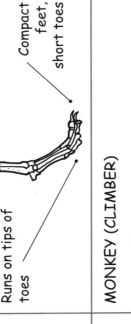

Shoulder joint moves mainly forward and back

Long leg bones

Compact feet, short toes

Runs on tips of toes

MONKEY (CLIMBER)

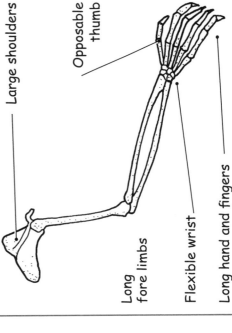

Large shoulders

Opposable thumb

Long fore limbs

Flexible wrist

Long hand and fingers

DOLPHIN (SWIMMER)

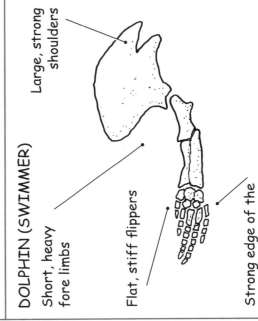

Large, strong shoulders

Short, heavy fore limbs

Flat, stiff flippers

Strong edge of the flipper cuts into the water

Linda Allison illustration.

HIND LIMBS: Some Known Bones

CHALLENGE:

FIND THE MODERN HIND LIMBS MOST LIKE THE MYSTERY FOSSIL BONES.

1. Get the card "Mystery Fossil Bones: Hind Limbs."

2. Compare the mystery hind limbs to the bones of modern animals shown on this page.

3. Which hind limbs do you think are most like the Mystery Fossil Bones?

_____

4. List the clues that helped you decide:

_____
_____
_____
_____

5. Based on the bone clues, what type of critter did the Mystery Fossil Bones belong to?

_____
_____
_____

Animals use their hind limbs or legs mostly for lo-comotion. Their front legs are often used for more specialized jobs like digging, paddling, or grabbing. Notice how different animals use their hind limbs:

● Wolves run on their long legs and compact feet. Strong hips support their body weight.
● Monkeys swing through tree branches with their long legs and grasping feet.
● Dolphins swim through water with their streamlined bodies. They lack legs.

Note how the shapes of the bones give clues to an animal's lifestyle.

DOLPHIN (SWIMMER)

Dolphins have no hind legs on the outside, but their skeletons have tiny, useless leg bones called vestiges.

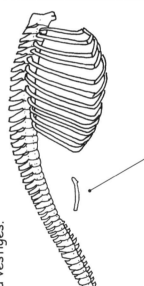

vestige leg bone

WOLF (RUNNER)

Strong, weight-bearing hips

Hip joint moves mainly forward and back

Long leg bones

Compact feet, short toes

Runs on tips of toes

MONKEY (CLIMBER)

Strong hip bones

Thumbs on feet

Long flexible toes

Hip joint has a wide range of motion

176

SPINE, NECK, & RIBS: Some Known Bones

CHALLENGE:

FIND THE MODERN SPINE, NECK & RIBS MOST LIKE THE MYSTERY FOSSIL BONES.

1. Get the card "Mystery Fossil Bones: Spine, Neck, & Ribs."

2. Compare the mystery spine, neck, & ribs to the bones of modern animals shown on this page.

3. Which spine, neck, & ribs do you think are most like the Mystery Fossil Bones?

4. List the clues that helped you decide:

_____

_____

_____

5. Based on the bone clues, what type of critter did the Mystery Fossil Bones belong to?

_____

_____

An animal's spine acts as a rod from which the other bones are hung. It is made of sections called vertebrae that form a strong, flexible support and provide anchors for the muscles. Ribs form a cage to protect heart and lungs. Notice how different animals use the spine in different ways:

- Wolves use their long necks to move the head for better sight, smell, and hearing.
- Monkeys travel through trees using their long tails for grabbing and balancing. Their flexible necks aid their keen vision.
- Dolphins swim using their powerful tails in an up and down motion to push through the water.

Note how the shapes of the bones give clues to an animal's lifestyle.

WOLF (RUNNER)

Long flexible neck

Deep narrow chest

MONKEY (CLIMBER)

Long flexible neck

Long balancing tail

Flexible grasping tail

DOLPHIN (SWIMMER)

Streamlined body shape

Short, stable neck

Long powerful tail

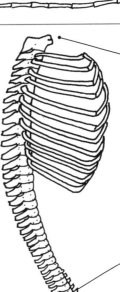

Linda Allison illustration.

Virus and the Whale: Exploring Evolution in Creatures Small and Large

2 Share Your Findings

Scientists seldom find complete skeletons. To learn more, they often must share information from other skeletons and from other dig sites.

a Partner with another group so Fore and Hind Limbs teams work together and Skull and Spine teams work together.

b Discuss your findings with partner group, and summarize them on one of the Sum It Up charts. See if you can come to an agreement about how your mystery animal moved, what it ate, and its habitat.

Sum It Up: Fore limbs and hind limbs

TEAM	ANALYSIS: Note your ideas about type of animal and its lifestyle
Fore Limbs	
Hind Limbs	
Both Teams	Which of the three modern animals might have a body part similar to this?
Summarize: How mystery animal moved, what it ate, and its habitat	

Sum It Up: Skull and spine

TEAM	ANALYSIS: Note your ideas about type of animal and its lifestyle
Skull & Teeth	
Spine, Neck, & Ribs	
Both Teams	Which of the three modern animals might have a body part similar to this?
Summarize: How mystery animal moved, what it ate, and its habitat	

3 Consider This

Philip Gingerich named the mystery bones *Rodhocetus*, the barely walking whale. This whale was found in an ancient shallow water habitat. After analyzing the bones, Dr. Gingerich believes that this animal lived both in and out of the water, and it was a meat eater that breathed air. How does your team's conclusion agree or disagree with Gingerich's analysis?

PART THREE
Creature features

When a new fossil skeleton is discovered, one of the first questions a scientist asks is what animal or family of animals it is similar to. Even a partial skeleton allows a paleontologist to identify features that match with other animals. By identifying similar features, scientists can sort which animals are closer relatives and where an animal fits into the overall scheme of living things. In the case of extinct animals, scientists can attempt to sort out if they are the ancestors of living animals, and how they evolved over time.

Take the next few minutes to consider the assembled *Rodhocetus* skeleton. Compare it to the skeleton of a modern dolphin (a type of small whale) and a modern shark. These two aquatic animals are similar in size but belong to different families. Check out the creature's features. Then decide if *Rodhocetus* is more closely related to a shark than to a dolphin.

Work with a partner
Each team will need:
- Creature Features: Comparing Anatomy sheet

1 Comparing Skeletons

Compare the three skeletons on the Creature Features sheet.

Would you say the modern whale or the modern shark is a closer relative of *Rodhocetus*? List the clues you noticed that helped you decide.

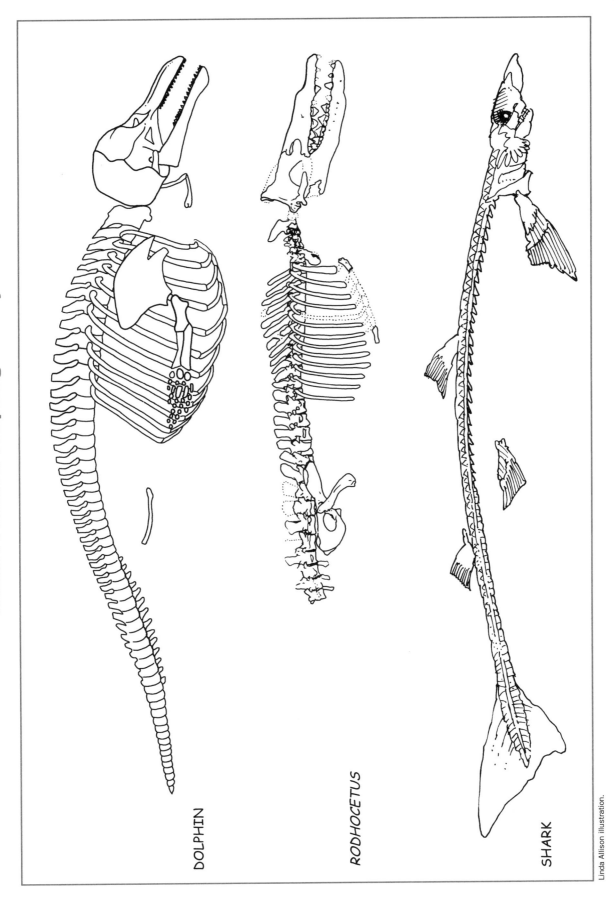

DOLPHIN

RODHOCETUS

SHARK

2 Consider This

Dr. Gingerich's original article about *Rodhocetus* did not include information about its tail. At that time, no tail for the fossil had been discovered. Try to predict what kind of tail belongs on this ancient whale. Consider the information given below. Then draw the tail on the partial *Rodhocetus* skeleton on the Creature Features: Comparing Anatomy sheet. Explain why you think this is the likely tail for this animal.

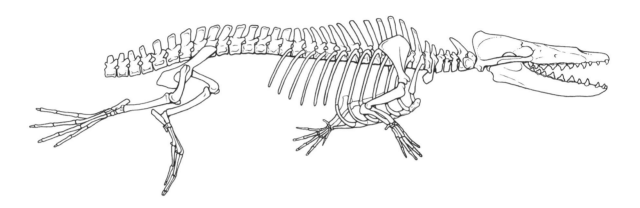

Fish and sharks have a finned tail that propels them forward by moving in a side-to-side fashion. Dolphins, like all whales, have a fluked tail that moves in an up-and-down motion.

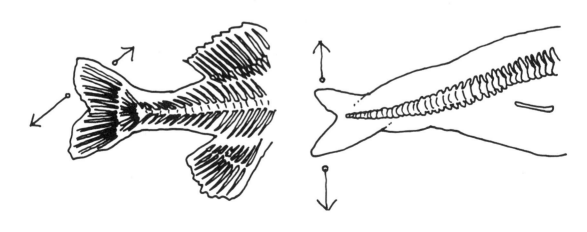

PART FOUR
Be a science reporter

Write a short news story about whales. Tell your readers about new fossil discoveries from Pakistan that suggest whales are related to animals that once lived on land. Based on what you have learned, explain how you think a whale could have an ancestor that lived on land.

P.S. Don't forget the headline.

Glossary

Accumulate

To collect or increase over a period of time. Changes in DNA, which are called mutations, accumulate over time at a steady rate.

Adaptation

An inherited change that improves an organism's chances of survival and of leaving more offspring. More of these individuals pass on their genes, so the inherited change becomes more common in the next generation. An example of adaptation is the whale's elongated and streamlined body, well suited for life in water.

Adaptive radiation

The evolution of one species into many, each adapted for a particular way of life. The 800 species of Hawaiian *Drosophila* that likely descended from a single female fly in the Hawaiian Islands is an extreme example of adaptive radiation. More than a dozen species of Darwin's finches that descended from one species in the Galápagos Islands is another example.

Adenine

One of the four chemical bases of nucleotides, the components of DNA that carry the code for making proteins, and, in turn, a complete organism. Adenine is often shortened to the letter A. See nucleotides, DNA.

AIDS

Acquired Immune Deficiency Syndrome. AIDS is a disease of the human immune system caused by infection with HIV, a virus that gradually destroys certain white blood cells, weakening the body's defenses against other diseases.

Ancestor

An individual from which another individual is descended, or an organism from which a species has evolved. Humans and chimpanzees are descended from a common ancestor that lived about 5–6 million years ago.

Ancestral

Belonging to past generations.

Antibiotic

A chemical produced primarily by some bacteria and fungi that kills or inhibits the growth of other bacteria, fungi, or other microorganisms. An example of an antibiotic is penicillin. The Penicillium mold releases the chemical penicillin, which humans use to kill a wide range of disease-causing bacteria.

Antibody

Substance produced by immune cells in the body in response to the presence of a virus or other invading organism. Antibodies provide immunity against viruses by binding to them and blocking infection.

Artiodactyl

Any hoofed mammal with an even number of toes on each foot. Artiodactyls include the pig, cow, sheep, deer, and hippopotamus. Fossil and DNA evidence both suggest that artiodactyls share an ancestor with whales and that the whale's closest living relative is the hippopotamus.

Attribute

A characteristic or quality of an organism, such as eye and hair color, that is controlled by genes.

Bacterium (plural, bacteria)

Any of a large group of one-celled organisms with a simple cell structure. Bacteria are found in most living things and in all the Earth's environments. They often live off waste nutrients of other organisms. While some cause diseases, others are beneficial. For example, trillions of bacteria live in our intestines, helping break down food. Bacteria are also a source of antibiotics, chemicals that kill other microorganisms that may cause illness.

Basilosaurus

A large extinct whale with a streamlined body 20 meters (60 feet) long, nostrils at the end of its snout rather than a blowhole, and tiny hind limbs, a reminder of its land-mammal ancestors. *Basilosaurus* is known from a large number of fossils. It lived in shallow seas around 40 million years ago. It is considered a relative, but not an ancestor, of modern whales.

Blowhole

The nostril or nostrils located at the top of the head of a whale that allows the exchange of air from the lungs.

Canine Teeth

Canine teeth are the pointed teeth next to the incisors in most mammals. They are suited for cutting and tearing meat.

Cell membrane

A structure that surrounds a cell and acts something like a window screen, controlling what passes in and out of the cell.

Chromosome

A tiny, threadlike structure in the nucleus of each cell that consists of DNA and protein and contains the recipe for making an organism.

Coevolution

Biological change in one species that is linked to change in another. An example of coevolution is the partnership that has evolved between farming ants and their fungus crop. Two other partners have also coevolved with the ant and the fungus: a crop pest that specializes in attacking the fungus crop, and bacteria that produce a special chemical that kills the pest but not the crop. By analyzing DNA, scientists are constructing the coevolutionary tree of the four partners. They have discovered that as one partner branched into new species, the other partners branched into new species as well.

Contagious

Transmitted from one organism to another by direct or indirect contact. The flu virus, for example, is transmitted by inhaling the virus from someone who coughs and sneezes. HIV is a virus that can be transmitted only through an exchange of blood or other bodily fluids.

Core sample

A sample of soil, rock, or ice collected by driving a hollow tube into undisturbed layers of the material and pulling out the sample or core. Scientists take core samples to study layers of material that are not visible from the surface. From animal bones, plant pollens, types of soils, and other deposits, scientists can gather clues about past climates and past life.

Cytosine

One of the four chemical bases of nucleotides, the components of DNA that carry the code for making proteins, and, in turn, a complete organism. Cytosine is often shortened to the letter C. See nucleotides, DNA.

Descendant

Any living organism related to one that lived in the past.

Diatom

A single-celled organism commonly found in lakes, rivers, ponds, oceans, and other wet environments. Diatoms use sunlight to turn carbon dioxide into food and oxygen. They are a primary source of food for many aquatic organisms, and the source of about half the oxygen on the planet. Diatoms have hard skeletons made of silica, a substance used to make glass.

Diverse

Many distinctive kinds. Hawaiian *Drosophila* are diverse, comprising over 800 different species.

DNA

Deoxyribonucleic acid, a double stranded molecule in the form of a twisted ladder that stores the recipe for making an organism. The rungs of the ladder are made up of four bases, called Adenine, Thymine, Guanine, and Cytosine. These bases, linked to a sugar and a phosphate group, make up chemical compounds called nucleotides. The sequence of nucleotides encodes the information needed to manufacture proteins, and, in turn, a complete organism. The sequence of nucleotides can

change or mutate, allowing life to evolve into new forms.

DNA fingerprinting

A technique for identifying individuals based on the uniqueness of their DNA pattern.

Enzyme

Any of the proteins produced in living cells that help speed chemical processes.

Evolution

A process that over many generations results in inherited changes in a population. In every species, individuals vary from one to another in their genetic make-up. These variations may be passed from one generation to the next in DNA or RNA. If individuals with a particular genetic trait survive longer and reproduce more than other members of the population, their genes will become more common over time. This process, operating over the last 3.7 billion years, has produced the amazing diversity of life on Earth.

Excavate

To dig carefully and methodically in order to uncover objects of scientific interest.

Extinction

The permanent disappearance of all members of a species or group of species.

Fertilization

The process in which a male sex cell (sperm) and a female sex cell (egg) unite to produce a cell that will grow into a new organism.

Fossil

Remains or traces (over 10,000 years old) of life.

Fossil record

A record of when different organisms lived in

the history of Earth. Information in the fossil record is based on dating layers of sedimentary rock to tell the age of a rock and the characteristic life forms in it.

Gene

A section of DNA that carries the chemical information needed to create a particular protein. Proteins are used to make and repair cells. Different proteins help determine how one type of cell differs from another, and ultimately, how one species of organism differs from another.

Genetic code

The sequence of nucleotides in DNA that specify which amino acids are to be used to build a protein. Each successive set of three nucleotides codes for one specific amino acid.

Guanine

One of the four chemical bases of nucleotides, the components of DNA that carry the code for making proteins, and, in turn, a complete organism. Guanine is often shortened to the letter G. See nucleotides, DNA.

HIV

Human Immunodeficiency Virus. HIV is a virus that causes the disease called AIDS. HIV invades the body's immune system, the very system that protects you from viruses and other invaders. Specifically, HIV targets the immune system's T cells.

Immune system

A system of cells and tissues in the body that work to protect the body against disease-causing agents, such as viruses, bacteria, parasites, and fungi.

Inheritance

The process by which traits or characteristics pass from parents to offspring through genes.

Leaf-cutter

A common name for some groups of ants that harvest fresh leaves for use in growing a fungus crop. These ants live in tropical forests in the Western Hemisphere. The scientific names of two kinds of leaf-cutter ants are *Atta* and *Acromyrmex*.

Molar

A large back tooth in mammals used for chewing and grinding.

Molecular clock

A means of determining the time that has elapsed since two species diverged, based on the assumption that DNA mutates at a steady rate. Scientists compare the same gene in individuals of different species and count the number of differences that have accumulated in the DNA. They multiply the number of differences by the average mutation rate for that gene and arrive at an estimate of the time at which the two species diverged. For example, humans and chimpanzees differ in about 100 nucleotides per 10,000 total in a gene called Xq13.3. The rate of mutation of that gene is about 1 every 50,000 years. For 100 differences to accumulate, humans and chimps must have branched apart about 5 million years ago.

Molecule

A small particle of matter made up of two or more atoms, held together in a chemical bond. An example of a molecule is DNA.

Mutation

A change in a portion of the DNA code. Mutations are responsible for providing genetic variations that allow living things to evolve into new forms. The DNA and RNA of all organisms, including humans, can mutate. Many mutations are completely harmless, because they strike parts of the DNA that are not essential in making proteins. Most other mutations produce defective proteins and may, in some cases, lead to the death of the organism carrying them. A few mutations may prove to be beneficial. A mutation that benefits an individual may, over generations, become a trait shared by all members of the species.

Mutual dependency

A partnership between two species that allows both to survive and benefit from the partnership. An example of mutual dependency is the partnership between leaf-cutter ants and the fungus they grow as a crop. The ants cannot survive without their fungus crop, which provides nutrient-rich food for the ants to eat. The fungus, in turn, cannot survive outside the ant nest. The fungus benefits from being groomed and cultivated by the ants. Mutual dependency is often the result of coevolution.

Natural selection

A process by which some individuals inherit traits that help them survive longer and produce more offspring than others. Over time, these traits become more common in the population. Scientists have been able to document the process of natural selection in many organisms. In a group of finches that live on one of the Galápagos Islands, finches with bigger beaks had a better chance of surviving a drought year, because they could eat the tougher seeds of a drought-tolerant plant. Within a few years of the drought, there were more larger-beaked finches in the population.

Neanderthal

An extinct species of human beings that once populated Europe and the Near East. Fossils of Neanderthals suggest they lived between about 200,000 years ago and 300,000 years ago.

Nucleotide

A chemical substance composed of one of several bases, Adenine, Thymine, Guanine, or Cytosine, linked to a sugar and a phosphate group. DNA is made up of chains of nucleotides. Nucleotides carry the codes that contain all of an organism's genetic information.

Nucleus

The cell's control center. The nucleus directs most of the cell's activities, including reproduction. It contains DNA and RNA, the genetic material of life.

Orangutan

A large species of ape found in the wild only in the forests of Borneo and Sumatra. It is one of the living mammals most closely related to humans, along with chimpanzees, bonobos, and gorillas.

Organism

Any living thing, such as a plant, animal, fungus, or bacterium. Scientists do not agree whether a virus is considered a living organism.

Pakicetus

The earliest known whale. *Pakicetus* lived about 48 million years ago. A well-preserved skull reveals ear bones that are unique to whales. Other features, such as canine teeth and a long snout with nostrils located at the tip, suggest it was a whale that lived on land and swam in shallow water.

Paleontologist

A scientist who studies ancient life through fossil evidence. By looking closely at fossil skeletons and other evidence of animal activity, and by comparing fossil bones with the bones of living animals, paleontologists try to work out what prehistoric creatures looked like, how they lived, and how they evolved. For example, paleontologist Philip Gingerich examines ancient whale fossils to answer questions such as which group of land mammals gave rise to whales, how and when the transition from land to sea took place, and what the transitional forms looked like.

Parasite

An organism that lives on or in another host organism in a way that can harm the host.

Petri dish

A round shallow dish used to grow bacteria and other microorganisms in a laboratory.

Pollen

A tough capsule that contains the male sex cells of a plant. Pollen is carried by wind or animals to other plants, where it fertilizes a female sex cell, forming a new plant.

Population

A group of individuals of the same species that live in a limited area and may interbreed with one another. A species is typically made up of several populations in different parts of a species' range.

Predator

An organism that hunts, kills, and eats other animals in order to survive.

Primate

A member of a group of mammals with a large brain, eyes facing forward, a shortened nose, and five digits on each of their hands and feet. Primates include humans, apes, monkeys, and lemurs.

Protein

A complex molecule made up of amino acids. Proteins are essential for life. They are the workhorses of the cell, carrying messages within and between cells, enabling chemical reactions to occur, and forming many of the structural components of living things.

Receptor

A cell structure or site that is located inside a cell or on a cell surface. Specific receptors are capable of binding with specific proteins and activating or inhibiting various cell functions. Viruses often enter cells by binding to receptors and fusing with the cell.

Replication

The process by which a virus copies itself. Viruses replicate by hijacking the cell's machinery. The cell then makes new viruses instead of its own products.

Reproductive success

The ability to produce more offspring than others in the population. It is an important component of natural selection and plays a primary role in sexual selection.

RNA

Ribonucleic acid, a single-stranded molecule that, like DNA, consists of long sequences of four different nucleotides. In living organisms, RNA is used to transmit the genetic code from DNA in the nucleus to the parts of the cell where proteins are built. Some viruses, such as HIV, have genetic information in the form of RNA rather than DNA.

Sediment

The material carried and deposited by rivers, lakes, seas, glaciers, and wind. Sediments deposited on a lake bottom or seabed often build up over millions of years to form sedimentary rock.

Selection

See natural selection, sexual selection.

Sexual selection

The process by which some individuals are more successful than others in mating and passing on their genes, because of characteristics that make them more attractive to a mate. If females are more attracted to one male than another, the more attractive male will father more offspring that will carry his genes. Likewise, females that choose better quality mates are likely to be more successful at producing more offspring and getting their genes into the next generation. Sexual selection explains why some organisms have features that serve no function other than to attract a mate: a peacock's elaborate tail, for example, or the elaborate songs and dances of male Hawaiian *Drosophila*.

Speciation

The processes by which new species are formed. If a population becomes isolated from the rest of its species by physical barriers, such as mountains, deserts, or stretches of open ocean, the isolated population will tend to evolve in a different direction from the rest of the species. It may eventually grow so distinct from other populations in the species that it would be unable to breed with them, thereby becoming a new species.

Species

A group of genetically similar organisms that can freely breed with one another, producing fertile young. In asexual or fossil organisms, species are defined solely in terms of genetic or physical similarity.

Strain

A group of organisms of the same kind that are descended from a common ancestor and share certain characteristics not typical of the group. Often used to refer to a group of viruses or bacteria that have the same line of ancestry and certain shared features. Examples are a vaccine-resistant strain of virus or a drug-resistant strain of bacteria.

Theory

An explanation that makes sense of a group of facts or events. A scientific theory is a comprehensive explanation of patterns found in nature, one that has been verified many times by different groups of scientists. Scientific theory gives rise to hypotheses that can be tested through experiments or observations. The atomic theory, the theory of relativity, the quantum theory, and the theory of evolution are all scientific theories that are so well established that they are the foundations on which modern science is built.

Thymine

One of the four chemical bases of nucleotides, the components of DNA that carry the code for making proteins, and, in turn, a complete organism. Thymine is often shortened to the letter T. See nucleotides, DNA.

Transition

A process of change involving passing from one state, stage, or form to another.

Vaccine

A preparation containing weakened or dead virus of the kind that causes a particular disease. A vaccine stimulates the immune system to produce antibodies against the disease.

Variation

Genetic differences in individuals that result in differences in their bodies and behavior.

Vertebrae

The bones of the spinal column.

Virologist

A scientist who studies viruses and the diseases they cause.

Virus

A minute organism that uses the cells of other organisms to reproduce. A virus consists of a small amount of genetic material (either DNA or RNA) inside a protein case.

X chromosome

One of the sex chromosomes. Humans have 23 pairs of chromosomes. Twenty-two of these pairs are the same in females and males. The last pair, the sex chromosomes, differ between the sexes. Females have two X chromosomes. Males have one X chromosome and one Y chromosome.

Resources

Selected resources on evolution for middle school teachers and youth leaders

Books About Evolution for Teachers and Youth Leaders

Burne, D. 2002. *Evolution: A beginner's guide to how living things adapt and survive.* New York: Dorling Kindersley.

Charlesworth, B., and D. Charlesworth. 2003. *Evolution: A very short introduction.* Oxford: Oxford University Press.

Gould, S. J., ed. 2001. *The book of life.* New York: W.W. Norton and Company.

Howard, J. 1982. *Darwin: A very short introduction.* Oxford: Oxford University Press.

Mayr, E. 2001. *What evolution is.* New York: Basic Books.

Weiner, J. 1994. *The beak of the finch: A story of evolution in our time.* New York: Vintage Books.

Zimmer, C. 2001. *Evolution, the triumph of an idea.* New York: Harper Collins Publishers.

Activities and Curricula About Evolution

Benz, R. 2000. *Ecology and evolution: Islands of change.* Arlington, VA: NSTA Press.

Lawrence Hall of Science. 2003. *Life through time.* Berkeley: University of California Press.

Stein, S. 1986. *The evolution book.* New York: Workman Publishing.

Resources on Teaching About Evolution

American Association for the Advancement of Science (AAAS). 1993. *Benchmarks for science literacy.* New York: Oxford University Press.

American Association for the Advancement of Science (AAAS). 2001. *Atlas of science literacy.* Washington, DC: American Association for the Advancement of Science.

Beardsley, P. M. 2004. Middle school student learning in evolution: Are current standards achievable? *The American Biology Teacher* 66: 604–612.

Bybee, R. W., ed. 2004. *Evolution in perspective: The science teacher's compendium.* Arlington, VA: NSTA Press.

Griffith, J. A., and S. K. Brem. 2004. Teaching evolutionary biology: Pressures, stress, and coping. *Journal of Research in Science Teaching* 41: 791–809.

National Academy of Sciences (NAS). 1998. *Teaching about evolution and the nature of science.* Washington, DC: National Academy Press.

National Research Council (NRC). 1996. *National science education standards.* Washington, DC: National Academy Press.

National Research Council (NRC). 2000. *Inquiry and the national science education standards: A guide for teaching and learning.* Washington, DC: National Academy Press.

Resources on Learning About Evolution

Bishop, B. A., and C. W. Anderson. 1990. Student conceptions of natural selection and its role in evolution. *Journal of Research in Science Teaching* 27: 415–427.

Brumby, M. N. 1979. Problems in learning the concept of natural selection. *Journal of Biological Education* 13: 119–122.

Clough, E. E., and C. Wood-Robinson. 1985. How secondary students interpret instances of biological adaptation. *Journal of Biological Education* 19: 125–130.

Evans, E. M. 2000. The emergence of beliefs about the origins of species in school-age children. *Merrill-Palmer Quarterly* 46: 221–254.

Evans, E. M. 2001. Cognitive and contextual factors in the emergence of diverse belief systems: Creation versus evolution. *Cognitive Psychology* 42: 217–266.

Lawson, A. E., and W. A. Worsnop. 1992. Learning about evolution and rejecting a belief in special creation: Effects of reflective reasoning skill, prior knowledge, prior belief, and religious commitment. *Journal of Research in Science Teaching* 29: 143–166.

Poling, D. A., and E. M. Evans. 2004. Are dinosaurs the rule or the exception? Developing concepts of death and extinction. *Cognitive Development* 19: 363–383.

Poling, D. A., and E. M. Evans. 2004. Religious belief, scientific expertise, and folk ecology. *Journal of Cognition and Culture: Studies in the Cognitive Anthropology of Science* 4: 485–524.

Resources on Evolution and Creationism

Jackson, D. F., E. C. Doster, L. Meadows, and T. Wood. 1995. Hearts and minds in the science classroom: The education of a confirmed evolutionist. *Journal of Research in Science Teaching* 32: 585–611.

National Academy of Sciences. 1999. *Science and creationism: A view from the National Academy of Sciences.* 2nd ed. Washington, DC: National Academy Press.

Numbers, R. L. 1992. *The creationists: The evolution of scientific creationism.* New York: Knopf.

Scott, E.C. 2005. *Evolution vs. creationism, an introduction.* Berkeley: University of California Press.

Skehan, J. W., and C. E. Nelson. 2000. *The creation controversy and the science classroom.* Arlington, VA: NSTA Press.

Web Resources about Evolution

http://evolution.berkeley.edu
Developed by the UC Berkeley Museum of Paleontology, this site supports a wealth of resources for teachers.

http://explore-evolution.unl.edu
This site provides information about the Explore Evolution project, with links to the museums where the Explore Evolution exhibit galleries are on display.

http://tolweb.org/tree
This site explores the National Science Foundation-funded Tree of Life Project.

www.nabt.org/sup/resources/ask.asp
This National Association of Biology Teachers site provides useful information about the teaching of evolution.

www.ncseweb.org
The home page for the National Center for Science Education, this site is a clearinghouse for information about the teaching of evolution.

www.nsta.org
This site provides information about resources available from the National Science Teachers Association.

www.nap.edu/books/0309053269/html/index.html
This site contains the entire text of the National Science Education Standards of the National Research Council.

www.pbs.org/wgbh/evolution
This PBS site includes video clips, interviews with scientists, and many other resources on evolution.

http://wonderwise.unl.edu
This site presents a series of nine multimedia science kits, each based on the research of a women scientist. These award-winning kits were developed by the same team that created the Explore Evolution project.

About the Authors

Linda Allison is an independent designer of learning experiences. She has created a wide range of educational tools and toys, including books, kits, and electronic media products for clients such as Texas Instruments, the Exploratorium, The Nature Company, Sony, Apple, and IBM. She has been a contributor to numerous award-winning projects, including the *New York Times* best-seller *Blood and Guts: A Working Guide to Your Own Insides.*

Judy Diamond, Ph.D. is the director of the Explore Evolution project and the Wonderwise, Women in Science learning series. Professor and curator at the University of Nebraska State Museum, she is the author with Alan B. Bond of *Kea, Bird of Paradox: The Evolution and Behavior of a New Zealand Parrot,* published by the University of California Press. She is also the author of *Practical Evaluation Guide, Tools for Museums and Other Informal Educational Settings,* published by AltaMira Press.

Sarah Disbrow, Ph.D. is an independent writer and editor of multimedia educational projects, museum exhibits, and trade book publications for Houghton Mifflin, Taunton Press, and Creative Homeowner Press. She is a contributing writer and an editor of the award-winning Wonderwise, Women in Science learning series and the editor of *Healthward Bound, An Owner's Manual for the Human Body,* produced to accompany the PBS program of the same name.

E. Margaret Evans, Ph.D. is a research investigator at the Center for Human Growth and Development at the University of Michigan. The Spencer Foundation and the National Science Foundation have funded her studies on the emergence of evolutionary concepts, including "Cognitive and Contextual Factors in the Emergence of Diverse Belief Systems: Creation Versus Evolution" published in *Cognitive Psychology.* Her dissertation was titled, "God or Darwin? The Development of Beliefs About the Origin of Species."

Carl Zimmer is the former senior editor of *Discover* magazine, a Guggenheim Foundation fellow, and the author of *Evolution, the Triumph of an Idea,* the companion volume to the PBS series broadcast in 2001. His latest book is *Soul Made Flesh: The Discovery of the Brain and How It Changed the World.*

Index

Index note: Entries beginning with a C indicate information contained in illustration section.

National Science Teachers Association